LIVING TOGETHER

LIVING APART

JEWS, CHRISTIANS, AND MUSLIMS
FROM THE ANCIENT TO THE MODERN WORLD

Edited by MICHAEL COOK, WILLIAM CHESTER JORDAN, AND PETER SCHÄFER

LIVING TOGETHER

LIVING APART

RETHINKING JEWISH-CHRISTIAN
RELATIONS IN THE MIDDLE AGES

Jonathan Elukin

PRINCETON UNIVERSITY PRESS

PRINCETON AND OXFORD

Library of Congress Cataloging-in-Publication Data

Elukin, Jonathan M., 1961–

Living together, living apart : rethinking Jewish-Christian relations
in the Middle Ages / Jonathan Elukin.

p.   cm. — (Jews, Chrisitans, and Muslims from the ancient to the modern world)

Includes bibliographical references and index.

ISBN-13: 978-0-691-11487-3 (hardcover : alk. paper)

ISBN-10: 0-691-11487-0 (hardcover : alk. paper)

1. Christianity and other religions—Judaism. 2. Judaism—Relations—
Christianity. 3. Jews—Europe—History—To 1500. 4. Social integration—Europe.
5. Jews—Persecutions—Europe. 6. Europe—Ethnic relations. I. Title.

BM535.E455 2007

261.2′60940902—dc22   2006022593

British Library Cataloging-in-Publication Data is available

CONTENTS

ACKNOWLEDGMENTS

It is a great pleasure to thank the many teachers, colleagues, and friends who helped me, directly and indirectly, write this book. Although this project was not based on my dissertation, it is in many ways the fruit of learning with my teachers at Princeton, particularly William Jordan. I am grateful as well to Anthony Grafton, Peter Brown, Natalie Zemon Davis, and Giles Constable. At the Jewish Theological Seminary, Benjamin Gampel, Stephen Garfinkel, Jay Harris, and Ismar Schorsch welcomed me into their classes. All of these scholars continue to be my teachers.

I have been fortunate to benefit from the wisdom, generosity, and support of many other colleagues, including Edward Peters, James Muldoon, Elisheva Carlebach, Ivan Marcus, Allison Coudert, Paul Freedman, Guy and Sarah Stroumsa, Elliot Horowitz, David Ruderman, Israel Yuval, Elchanan Reiner, Daniel Schwartz, Isaiah Gafni, John Van Engen, Robert Bartlett, Miri Rubin, Robert Stacey, Teo Ruiz, Esther Cohen, and Patrick Geary. Many of these distinguished scholars might disagree with aspects of this project, but I hope they recognize that I draw on their work for inspiration.

Brigitta van Rheinburg helped me to shape the book and was an advocate for it at Princeton Press. Nathaniel Carr and Jack Rummel took great care of the manuscript. The two readers for the press challenged me and helped to make the book better even if they did not accept all its arguments. I am responsible, of course, for any errors.

I am fortunate indeed in my friends in Chicago, Princeton, Washington D.C., Philadelphia, Jerusalem, and now in Hartford, particularly David Hahn and Iddo Landau whose friendships are great blessings. Trinity College provided sabbatical time and research support for the completion of this project. The intellectual companionship of my friends, colleagues, and students at Trinity energized my thinking about this book.

A few lines in an acknowledgment are meager thanks for my parents, Harriet and Edward Elukin, whose love and support are steadfast. My wife's parents and her siblings have graciously welcomed me into a larger family.

There are many heroes in my family, such as my father and father-in-law, who fought this country's wars. This book is dedicated to one of them: my mother's favorite uncle, a physician who died as a young man defending his country many years ago.

As I say to myself every Friday night, my wife Tracy is truly a woman of valor. She has made a home and a life for us where I can write and teach. In studying the past, I can only guess at what Jews and Christians in medieval Europe thought about the future. When I think about my beloved son, Benjamin, I see a future of great hope.

J.E.
April 12, 2006 (Passover Eve)

LIVING TOGETHER

LIVING APART

INTRODUCTION

AFTER TEACHING a course on relations between Jews and Christians in the Middle Ages for several years, I noticed a recurring query posed by many students. How did Jewish communities continue to survive in Europe despite facing what seemed to be endless persecution, violence, and expulsion? A fundamental question to be sure, but one to which I did not have a ready answer. My own work on the conversion of Jews to Christianity grew out of the sense that relations between Christians and Jews were driven largely by Christian antagonism to Jews.[1] In trying to resolve the paradox of persecution and survival with my students, I felt I was missing the real significance of Jewish-Christian relations. I began to think of ways to respond to this larger issue of the long-term resilience of Jewish-Christian relations in medieval Europe. The result is this book, really an extended essay, which tries to reorient our understanding of the meaning of the history of Jews in medieval Europe.

For too long scholars have tried to find that meaning in the nature of Jewish suffering in the Middle Ages. Their conclusions reflect the rhetoric of dispersion and suffering embedded in classical Jewish and Christian thought. Jews themselves drew on the biblical tradition of exile to understand their condition under Christian rule. Even before Christianity's advent as the official religion of the Roman world in the fourth century, Jews were used to the idea of living in a diaspora. Whether Jews experienced life outside the Land of Israel exclusively as suffering and trial has recently been challenged.[2] Whatever the truth of the long-term experience, the *idea* of suffering in the dias-

1

pora quickly became embedded in liturgical and other forms of Jewish religious culture after the destruction of the Second Temple in 70 CE.

Yosef Yerushalmi has analyzed how early rabbinic tradition and later medieval Jewish authors of prayers and chronicles created a kind of endless loop of Jewish suffering. They associated contemporary persecutions with traditional dates of traumatic destruction such as the fall of the Temple. In this foreshortened and essentially ahistorical sense of the Jewish past, all episodes of suffering and persecution were essentially the same; they all derived from God's testing and chastisement of the Jews.[3]

Even after the Enlightenment and the growing acceptance of Jews into European society, these liturgical memories of persecution and suffering provided the mental parameters of Jewish historiography of the nineteenth century. Jewish scholars of the *Wissenschaft des Judentums* movement—following in the footsteps of earlier eighteenth-century enlightened scholars (*maskilim*)—heroically tried to assert the value and vitality of the Jewish past using the tools of modern historical research. However, they could not (or did not wish to) escape fully from the narrative of persecution and suffering.[4] The suffering of the Jews—or perhaps the survival of the Jews despite great suffering—was a way for nineteenth-century historians such as Heinrich Graetz to strengthen a sense of Jewish community as well as signal a break with the past. History could show that the emancipation of the Jews in modern Europe, albeit imperfect, offered an escape from the persecution and obscurantism of the past. At the same time, it was thought the Jews played a crucial role as messengers of ethical monotheism and were full participants in the history of Europe. Ultimately, Zionist historians disdained much of the Jewish past as a history of persecutions and diaspora, or recast it in more palatable terms emphasizing the survival of the Jewish nation.

The dispersion and suffering of the Jews made perfect sense to Christians. The Church Fathers saw the dispersion of Jews as justified punishment for the Jewish rejection of Jesus. Moreover, the denigration of Jews was cited by generations of Christian polemicists as proof of God's rejection of Israel. The Augustinian model of toleration of the Jews, which would prove so important to Jewish-Christian relations, was still based on the divinely ordained subservience of Jews to Christians.

Protestant historians who studied the Middle Ages, building on humanist prejudices of the Renaissance, saw medieval Catholic Europe in general as a superstitious and violent deformation of true Christianity. The medieval treatment of Jews fit what they thought was a cruel and corrupt church. Once Protestants moved beyond Lutheran anger at the Jews for not converting, eighteenth-century scholars such as Jacques Basnage articulated a more sympathetic view of Jewish history.[5] The Catholic historical tradition answered the Protestant challenge with its own romanticized vision of a hierarchical and natural medieval world. In this vision, Jews were pushed to the margins and fixed in historical sensibility as usurers who provoked what violence was visited on them. This was in many ways an echo of the original Christian sense that Jewish history as a coherent, meaningful narrative ended after the incarnation.[6]

As the professional academic study of the Middle Ages evolved in the still Christian universities of Europe and America, the experience of Jews faded as a central concern of academic scholarship. Without theology driving interest in Jews, historians of the Middle Ages seemed largely indifferent to the history of Jews in medieval Europe. The work of Jewish historians was largely ignored. When Jews were mentioned,

what mattered was their persecution or their recurring historical role as moneylenders.[7]

More recently, the fate of Jews, as well as that of other minorities in the Middle Ages, has become a central concern of the
historical discipline. The Holocaust, of course, challenged
scholars to look for what went so tragically wrong in Christian
attitudes toward Jews. The rise of ethnic history in general has
encouraged a flowering of scholarship on the Jewish past and
accorded it a new standing in the academy. The new attention
to minorities in medieval society may be due as well to the
influence of multiculturalism and postmodernity's interest in
the bizarre and marginal.[8] Whatever the motivation behind this
postwar scholarship, the larger narrative structure, emphasizing fundamental and constant persecution, echoes earlier approaches to Jewish history.

The historiographical fulcrum for much of the recent work
on the treatment of Jews is the claim that twelfth-century Europe became a "persecuting society."[9] The treatment of Jews in
the medieval past thus ominously signals the fundamentally
intolerant character of European states and Christian culture.
Although a suggestive characterization, this interpretation
reads back into medieval history the anachronistic power and
efficacy of the twentieth-century totalitarian state.[10]

Seeing medieval Europe as a persecuting society obscures
the complexities of the Middle Ages and reduces the Jewish
experience to a one-dimensional narrative of victimization.
Making Jews the "Other" of medieval Europe, a group singled
out for marginalization and persecution, creates arbitrary categories that do not reflect medieval realities. Indeed, does this
increasingly popular term actually help us to understand the
dynamic between groups in medieval Europe? What does such
a label mean in the complicated world of the Middle Ages?
As Paul Freedman has recently observed: "The Middle Ages

certainly created a panoply of mistrusted and persecuted enemies—Saracens, Jews, lepers, heretics, apocalyptic peoples. But
the very heterogeneity and proliferation of such despised peoples calls into question how 'the Other' is to be used as a theo-
rizing tool."[11] Even if such distinctions shaped thinking about
nonnormative groups, the very idea of an "Other," as Freedman notes, suggests that "elite society is presented as unani-
mously and unquestioningly determined to push a variety of
feared or despised peoples to the margins of the human."[12] This
Manichean vision of medieval Europe ignores the complexities,
paradoxes, and tensions within elite society. Moreover, it en-
courages a scholarly emphasis only on the persecution of Jews.

This one-dimensional interest in Jewish suffering has not
gone unchallenged. Salo Baron famously called in the first half
of the twentieth century for scholars to move beyond seeing
Jewish history as only the story of persecution. However, his
challenge was not really taken up by scholars in any systematic
fashion.[13] Recently, John Hood's book on Aquinas, Ivan Marcus's treatment of education and cultural assimilation, Robert
Lerner's study of positive apocalyptic attitudes toward the Jews,
and Johannes Heil's rereading of 1096 all suggest that persecution is not the entire story of Jewish history in the Middle
Ages.[14] Violence against Jews, as we have been reminded, was
contingent on local conditions and not the result of unchanging hatred or an irrational structure of medieval society.[15]

I think that it is time for an attempt to rethink Jewish history
of the medieval and early modern period along these new lines.
Indeed, as I argue below, the older histories have inadvertently
distorted or at least obscured our ability to see a fuller range
of Jewish experiences in the Middle Ages. Michael André Bernstein has recently cautioned against the practice of writing history with the end in sight.[16] As a result, we have lost sight of
what is most important about the Jewish past in medieval

Europe. Instead of persecution and suffering, it is more important to understand how and why Jews survived in societies whose dominant theology increasingly cast them in the role of deicides.[17]

The categories of persecution and tolerance focus our attention on only one aspect of relations between Christians and Jews. We should look to see how Jews and Christians of medieval Europe engaged each other across a larger spectrum of relationships and experiences. How did Jews *live* in medieval European society as it evolved? It is this task that the first two chapters of the book take up. I believe that we will find that Jews of the Early and High Middle Ages were deeply integrated into the rhythms of their local worlds. Moreover, they faced many of the same challenges as their Christian neighbors. The variety and dynamism of medieval Christianity created a society in which the Jews were not alien interlopers facing a uniformly antagonistic Christian world. In chapters 3 and 4 I will discuss how that rootedness was expressed in cultural and social integration into Christian Europe.

The integration of Jews into medieval society and culture makes it easier to understand how Jews could analyze, contextualize, and hope to manage the violence that did erupt. Their rootedness in local societies and their familiarity with local patterns gave them skills that would help them survive or at least to take actions that they thought would protect them. As I will argue in chapter 5, the nature of medieval violence itself affected how Jews perceived the violence directed against them. Jews were used to seeing aggression against them in the larger context of medieval social conflict that drew in many other groups. Jews were aware of the high levels of physical and rhetorical violence displayed against many groups and individuals in European societies. Jews were not singled out in medieval societies as the preferred target of violence. Moreover, the level of violence against Jews—either oppressive laws, outright attacks, paranoid accusations, or expulsions—were essentially transitory and contingent events that did not fundamentally destroy the modus vivendi between most Christians and Jews of the time. The transient nature of the violence gave Jews a sense of fundamental stability and security. This discussion should help us rethink our reliance on the idea of the Middle Ages as a "persecuting society" in which continuous repression of Jews was a fundamental part of medieval culture.

From 1290 onward, expulsion was the primary form of aggressive governmental action taken against Jews of medieval Europe. Jews were pushed out of most Western European countries (with the exception of Italy and parts of Germany) by the second half of the sixteenth century. They found refuge in Ottoman Turkey but mostly in Christian Poland and Lithuania. Such a widespread movement against Jewish communities, combined with perceptions of the Middle Ages as a crucible of persecution, makes it is easy to see how the expulsions have been cast as the inevitable culmination of anti-Jewish feeling. In chapter 6, I hope to challenge this underlying sensibility and emphasize that the actions and policies that led to final expulsions from England, France, and Spain—or temporary expulsions in Italy and Germany—should not be considered inevitable. Such an argument is not naïve. The expulsions happened. Christians planned and executed them, and many Jews suffered from the dislocation and the associated violence. But we are used to seeing them as completed actions, linked together by a common culture of anti-Jewish sentiment, not the unfolding of uncertain, contingent, and separate events that did not necessarily reflect the sentiment of most Christians.

Indeed, the "end" of the Middle Ages, or rather the perceived division between the Middle Ages and the Reformation/Renaissance that historians have constructed, has played a large role in shaping our understanding of Jewish-Christian relations. In chapter 6 I will also argue that the traditional conception of the Renaissance and Reformation has distracted us from the fundamental continuities in relations between Christians and Jews into the sixteenth century. The second half of the sixteenth century saw the end of the expulsions and the beginning of the readmission of Jews to many of the localities from which they had been expelled. The experience of Jews in Italy and Germany at this time suggests many continuities with medieval conditions. It may get us closer to the lived experience of Jews during that time to see the expulsions not as the conclusion of the medieval phase of Jewish-Christian relations but rather as periodic breaks in a larger trajectory of generally stable Jewish-Christian relations that survived into the early modern period.

Ultimately, this book is an attempt to read medieval Jewish history against the grain. The narrative of Jewish history in the Middle Ages, whether consciously constructed or not, has been one of rising persecution that ends in expulsion. This is my hope: "The historian restores history to the complex situation which prevailed when it was still in the course of being decided. He makes it into the present once more, reviving its acute alternatives. In the true sense of the word, he makes it happen again, that is, he has it decided again. He dissolves the content, the product, the form of the completed work or the done deed, at the same time appealing to the will, to the living power of decision, out of which these works and deeds grow."[18]

I am acutely aware that I am dependent on the work of others for many aspects of this book, particularly the later surveys of Italy, Spain, Germany, and France. The interior culture of the Jews is also neglected in this book. (The growing literature on various aspects of medieval Jewish culture would, of course, help make the case that Jews found Europe to be more or less congenial since a flourishing Jewish civilization arose even under the relative constraints of Christian Europe.) In order to construct a comprehensive argument about the nature of Jewish-Christian relations in the medieval and early modern west, I have strayed far from my own limited areas of expertise. Even where I disagree with the interpretations of others, it is their focused scholarly efforts that make a work of synthesis possible.

Others will likely disagree, but the fundamental truth or meaning of Jewish history in the Middle Ages—if we are right to apply such a term as meaning—is the continuity of relatively stable relations between Jews and Christians. Readers will notice quickly, and many will no doubt criticize, my inattention to the cultural expressions of medieval anti-Judaism. It is clear that during the twelfth and thirteenth century and continuing into the later Middle Ages, images and language about Jews was increasingly aggressive and derogatory. (At points the cultural pressures could be translated into policy—particularly the actions of pious kings under the influence of mendicant anti-Judaism). However, documenting this hatred still leaves the question of how Jewish communities survived in such a cultural atmosphere. How do we account for the relatively limited violence given the viciousness of the images? The cultural expressions of anti-Judaism may thus be giving us a distorted vision of the total experience of medieval Jews.

The objection might be raised as well that examples of good relations between Jews and Christians should not outweigh the general anti-Judaism of the culture in a general assessment of the medieval Jewish experience. I see the problem, or the potential solution, differently. It is difficult to imagine that relatively open relations between Jews and Christians could have formed unless social norms allowed room for such contacts. If mutual antipathy was truly and constantly dominant, how could such relations have evolved? Peaceful relations between Jews and Christians did not mean that antagonism did not

exist, but that animus was not a constant feature of social relations—or at least that hatred and suspicion was controlled. In
this sense, "normal" interactions between Jews and Christians
may not be exceptions but indications of much greater fluidity
in medieval Christian mentalities about Jews as well as Jewish
reactions to Christians.

By the sixteenth century, attempts to short-circuit relations
between Christians and Jews had apparently succeeded. But
with the decision to readmit or to stop expelling the Jews, European authorities allowed a return to a pattern of relations established in the Middle Ages. What requires recognition and explanation is not the actions of medieval governments or violent
groups of Christians, which can often be explained in terms of
situational policies and pressures, but the resilience of the
modus vivendi forged in the Middle Ages between European
Jews and Christians. We need to understand how Jews and
Christians lived together while still living apart.

⚘ 1 ⚘

From Late Antiquity to the
Early Middle Ages

THE PERIOD from the fifth century and the first incursions of non-Roman peoples into the territories of the empire to the end of the Carolingian hegemony in the ninth century is usually seen as a period of relative security and tolerance for Jews.[1] By trying to explain the tolerance, scholars have emphasized how unusual it is and how it degraded into more "normal" persecutorial relations after the year 1000. This division of medieval history between an early period of tolerance and a later period of persecution prevents us from seeing the dynamic nature of the Jewish experience before and after the year 1000. Moreover, asserting the tolerant quality of Christian treatment of Jews in early medieval Europe has had the unfortunate effect of setting in apparently higher relief the persecutions of Jews in the centuries following the First Crusade. The experience of Jews in the early years of the medieval period was certainly more complicated than any clear-cut opposition of tolerance or persecution. Jews and Christians of early medieval Europe engaged each other across a larger spectrum of relationships and experiences. Jews were deeply integrated into the rhythms of these local worlds, living as natural participants in the culture, politics, and societies of early medieval Europe.

Our search for usable evidence to illustrate the complicated dynamic between Christians and Jews will lead us quickly from the island of Minorca in the fifth century, to the towns of sixth and seventh century Gaul, to the divided society of Italy in the seventh century, and to Visigothic Spain of the same period. These variegated societies set the boundaries for the experiences of Jews in Europe until the eighth and ninth centuries, when the descendants of Charles Martel—in particular—Charlemagne, gave some kind of unified religious and political identity to Western Europe.

The first "micro-Christendom" where we can see something of how Jews and Christians lived together is the island of Minorca in the early part of the fifth century. Jews could look back on centuries of continuous settlement in many of the towns and cities of Roman territory in Spain, France, and Italy.[2] Even under the increasingly harsh decrees of Christianized imperial legislation, the Jews still had been guaranteed the protection of Roman laws to practice their religion.[3] Early Christian theology, articulated by Augustine, reinforced this attitude in Christian terms by asserting that Jews must be allowed a place in Christian society since they helped prove the antiquity of Christianity and the authenticity of Scripture. The transition to a Christian empire did not radically alter the restrictive toleration of Jews.[4] The laws emphasized the second-class status of Jews by prohibiting the conversion of Christians to Judaism. They limited contact between Jews and Christians by prohibiting mixed marriages and tried to prevent Christians from attending Jewish religious rituals. The social and physical sphere of Jews was restricted by the prohibition on the construction of new synagogues. At the same time, and this is the tension that would animate much of church law on Jews for the rest of the Middle Ages, Jews were guaranteed the freedom to practice their religion, even if that freedom was granted in denigrating and abusive language.[5]

The fifth century presented challenges to Jews and Christians alike. Christians were still struggling to shape their own religious identity. We are still not sure what being Christian meant to the mass of people of these largely agricultural societies.[6] Moreover, those people who identified in some way as Christian were constantly subjecting their own religious identities to challenges and redefinitions. Catholics struggled violently among themselves and with Arians, Manichees, Donatists, Pelagians, and others as each group sought to define itself as orthodox and others as heretical.[7] This was not an irenic, unified Christian world; it became even more complicated when the Roman Empire, increasingly Christianized after the conversion of Constantine in 337, was reshaped in the fifth century by massive incursions of ethnically non-Roman tribes from the north and east of Europe. These warrior groups, many already culturally Romanized, created new elites, ethnic identities, and sources of power in the lands of the empire.[8] Even if the "fall" of Rome was really a "transformation," the accompanying tragedies of wars, famines, and plagues shaped a fifth century in which Jews and Christians alike faced very uncertain futures.

It is against this background that we find the letter of a Christian cleric, Severus, describing the forced conversion of some members of the local Jewish community in Minorca to Christianity in 418, following the arrival of the relics of St. Stephen on the island.[9] As we will see throughout the period, Jews maintained a range of personal relationships with Christians. And even under the open violence of radicalized Christians, Jews responded as fellow participants in a local culture rather than as an isolated and persecuted minority. Jews were firmly established in the local society of Minorca before the conflict with Christians began. In apparent opposition to or disregard of some of the provisions of the Theodosian Code, Jewish leaders dominated the political life of the island, with one serving as *defensor*—essentially governor—for the island. According to Severus,

> the Jewish people relied particularly on the influence and knowledge of a certain Theodorus, who was pre-eminent in both wealth and worldly honour not only among the Jews, but also among the Christians of that town [Magona]. (2) Among the Jews he was a teacher of the Law and, if I may use their own phrase, the Father of Fathers. (3) In the town, on the other hand, he had already fulfilled all the duties of the town council and served as *defensor*, and even now he is considered the *patronus* of his fellow citizens.[10]

As Severus looked back to a time before the period of conflict, he recalls that relations between Jews and Christians had been marked by a warm civility. "In the end," he writes, "even the obligation of greeting one another was suddenly broken off, and not only was our old habit of easy acquaintance disrupted, but the sinful appearance of our longstanding affection was transformed into temporary hatred, though for love of eternal salvation."[11] The bishop could not create a false history of the island where Christians had always been antagonistic toward Jews. Some kind of social peace had prevailed before the arrival of the relics. That was what was so galling to the radicalized cleric.

In addition to the normal bonds of neighbors, Jews and Christians shared a common liturgical culture based on Scripture. This connection was apparent even to Severus when he remembered the preliminaries to a debate between the two communities. As Jews and Christians marched together to a synagogue for a public debate, "along the way we began to sing a hymn to Christ in our abundance of joy. Moreover, the psalm was 'Their memory has perished with a crash and the Lord endures forever' [Ps. 9:7–8], and the throng of Jews also began to sing it with a wondrous sweetness."[12] Did the two groups use the same tune, or did Jews sing in Latin or Greek with the Christians? Or did the Jews recognize the psalm and begin a counter singing in Hebrew? Whatever the case, Severus recognized that the intimate sounds of liturgical music and a shared if contested Scripture connected the two communities.

As the letter opens, Severus's attempts at converting Jews had been going on for well over a year when the relics of Stephen arrived.[13] Nevertheless, Severus remembered the appearance of the relics as the event that galvanized his followers. The physical reminders of holy men and women had become the focus of Christian religiosity during the past two centuries. They provided Christians with a locus for prayers and miracles of healing. Moreover, early Christian bishops cemented their control over local communities by orchestrating the discovery and translation of the relics of local holy men and women.[14]

The relics may have inspired Severus, but at least in Minorca, there was no sudden conversion of the mass of Jews; it took an enormous amount of pressure to bring Jews to the baptismal font. Jews had to make difficult choices when faced with radicalized Christian violence. Some Jews followed the great men of the community into the Christian fold.[15] Their leadership was not enough to persuade all the Jews. Severus describes stubborn resistance: some individuals took to the caves or held out in small groups.[16] Severus remembers that Jews invoked the Maccabees as models for their struggles against the Christians.[17] Individual actions under these conditions could never be predicted. Severus records that "a certain Jewish woman (by God's arrangement, I suppose) acted recklessly, and doubtless to rouse our people from their gentleness, began to throw huge stones down on us from a higher spot."[18] An unknown number seemed to wait out the explosion of Christian radicalism and perhaps later returned to a more openly Jewish identity. The Jewish response reinforces the impression that this was a dynamic situation where individual actions and perceptions of Jewish power and security affected the outcome of the confrontation.

Whether by miracles or intimidation, some significant number (more than five hundred if we believe Severus's account) of Jews ultimately converted. The editor's scholarly detective work has revealed, however, that Severus's victory was not as absolute as he wished to present it. In fact, even in 419—after the conversions—the Christians were still faced with a resilient Jewish community.[19] The Jews of Minorca had not been destroyed. Their great men still occupied positions of influence and power—albeit some as newly converted Christians.[20] Perhaps Severus had exaggerated the number of conversions in the first place. More likely, some conversions had been temporary measures, and part of the Jewish community reconstituted itself after the first wave of Christian intimidation.

It is important to remember that what Jews faced in Minorca was not unusual for other people throughout the Romanized world of the early fifth century. As the Huns, Vandals, Goths, and then other migrant peoples disrupted the societies of the empire, many localities would have seen groups persecuted or relocated. In this new environment when the reach of the central authorities in Rome and Constantinople had begun to weaken, many populations had to negotiate for their security and survival with local leaders. Under these conditions, traditional definitions of identity and loyalty would have come under great pressure. Jews of Minorca, at least, could weather flashes of Christian violence and extremism because their roots in the locality went very deep.

In addition, their connections to neighboring communities gave Jews refuges in towns that had not seen such outbreaks of Christian violence. As Severus recalled,

> there still remained two women who refused to race to the fragrance of Christ's unguents: the wife of that Innocentius whom we mentioned above, along with her sister, a widow of excellent reputation. (2) Yet the moment she learned that her sister's husband, Innocentius, had been converted, she boarded ship. We not only permitted her to do this, we even encouraged her, because she could not be turned to faith in Christ by either words or miracles.[21]

The Mediterranean world still offered many safe harbors for Jews.

⁂

We can see many of the same patterns in Jewish-Christian relations slightly later in Merovingian Gaul. By the fifth century, the barbarian people known to the Romans as Franks had come to dominate these territories after many years of jockeying for power with local Roman leaders as well as other non-Roman peoples. By the end of the fifth century, a more or less unified kingdom was established when the great king Clovis, who had embraced Christianity (Catholicism) before a crucial battle, founded his own dynasty. At least that is the image we have from Gregory of Tours who recounts Clovis's exploits in his *Ten Books of History*, written in the sixth century.[22]

In many ways, Christianity was a newer religion for the Franks than for the more Romanized people in Minorca. It is impossible to know what kind of Christianity or Christian identity was formed in the melting pot of Arians, Catholics, Romans, and Franks after Clovis converted. We really only have Gregory's impression of how the Frankish military suddenly adopted Christianity after Clovis's conversion.[23] For the larger Frankish society of Clovis's day, which included Roman natives of Gaul, and even of Gregory's a generation later, we know very little about the nature of Christianization.

We cannot assume that Gregory's own intense and internalized sense of Christian identity was diffused throughout all levels of Merovingian society. His repeated insistence on the efficacy of relics and saints' tombs to perform miracles suggests that he felt he had to convince a skeptical audience about the new religious economy of Christian healing and salvation. He knew there were alternatives to the messages of Christian bishops. The common peasants were easily deceived, as Gregory delights in telling us, by fake miracle workers and prophets:

> That same year there appeared in Tours a man called Desiderius, who gave it out that he was a very important person, pretending that he was able to work miracles. He boasted that messengers journeyed to and fro between himself and the Apostles Peter and Paul. I myself was not there, so the country folk flocked to him in crowds, bringing with them the blind and infirm. He set out to deceive them by the false art of necromancy, rather than to cure them by God's grace.[24]

Gregory was not too sanguine either about the capacity of more educated people to become true Christians. Even one of his own priests, he recounts, could not bring himself to believe in bodily resurrection.[25] At the very least, we have to be open to the possibility of a rather differentiated population with a variety of ideas about what constituted normative or correct Christianity. The world of Gregory of Tours and the Franks of early medieval Gaul is not then necessarily a monolithic Christian community in which all people identifying as Christian shared the same self-identity, or, for that matter, beliefs about

Jews. It is unlikely then that Jews would have felt themselves facing a homogenous Christian population. (Emphasizing the varied nature of Christian belief and habits in Gaul is not the same as insisting on a very thin layer of Christian belief covering a deep rooted and resilient paganism.)[26]

When Gregory does take note of Jews—which he does infrequently—what is apparent is how they seem a natural part of the local society of Merovingian Gaul. Even in one of Gregory's lives of the local martyrs, he gives us a picture of Jewish and Christian children studying and learning the alphabet together.[27] The bonds to a particular locality would have been strong in what essentially remained an agricultural society. Even Jews in towns would have connections or interests in the agricultural hinterland on which their lives depended and from which real wealth and status in society derived. These Jews and Christians met on all social levels.[28] That interaction often provoked Christian clerics. Gregory, for example, chastised Christians who consulted Jewish doctors. He recounted one story of a man who rejected the healing offered at a proper Christian shrine and

> went off home and consulted a Jew, who bled his shoulders with cupping-glasses, the effect of which was supposed to be that his sight would improve. As soon as the blood had been drawn off, Leunast became as blind as he had been before. He thereupon returned once more to the holy shrine. There he stayed for a long time, but he never recovered his vision . . . Leunast would have retained his health, if he had not sought the help of a Jew after he had received God's grace.[29]

Jews were an accepted part of the landscape of these Merovingian towns, even if Gregory despaired about their role in society.

On a larger public scale, Jews were recognized as participants in local political culture. They felt connected to local rulers and expressed their loyalties in public. When, for example, King Guntram entered the town of Orleans in 585 Jews were part of the multiethnic crowd that acclaimed him. Gregory writes:

> The day of his entry into Orleans was the feast of Saint Martin, that is 4 July. A vast crowd of citizens came out to meet him, carrying flags and banners, and singing songs in his praise. The speech of the Syrians contrasted sharply with that of those using Gallo-Roman and again with that of the Jews, as they each sang his praises in their own tongue. "Long live the King!" they all shouted. "May he continue to reign over his peoples for more years than we can count! Let all peoples continue to worship you and bow the knee before you and submit to your rule!"[30]

Local Jews seem to behave and are recognized as a normal part of the population. Gregory later praises Guntram for not allowing the blandishments of Jews to influence him in the rebuilding of a synagogue. However, the portrait Gregory paints of the king's entry suggests that Jews—like everyone else—were more interested in the general order and stability that the king might bring than in any specific favors he might grant them.

Gregory shows us how Jews moved among all levels of Merovingian society. A Jew named Priscus, for example, had regular contact with King Chilperic (d. 584) as a purchasing agent. Their relationship extended to a kind of bantering debate over religion. As Gregory tells it, he witnessed a rather frank exchange between Priscus and the king: "The Jew replied: 'God has no need of a Son. He has not provided himself with a Son and He does not brook any consort in His Kingdom, for He said through the mouth of Moses: See now that I, even I, am he, and there is no god with me.' "[31] They continued to trade scriptural quotations until the Jew asks in disbelief, "How should God be made man, or be born of woman, or submit to stripes, or be condemned to death?"[32] At this the king falls silent and Gregory takes over with a stream of scriptural citations designed to persuade the Jew. It was to no avail, as Gregory admitted: "Despite all my arguments, this wretched Jew felt no remorse and showed no signs of believing me. Instead he just stood there in silence."[33]

The dispute takes its place in Gregory's history with the other theological conflicts that mar Merovingian society. These theological arguments never seem to change anyone's mind in Gregory's narrative, whether they involve Arians, proto-Saddu-cees, or Jews.[34] Gregory included them perhaps to show what a bishop should do and what scriptural resources he had at his disposal. The reluctance of the Jew to admit the truth is another part of the scenery of a Gaul disturbed, corrupt, and violent. Even Chilperic's reported forced conversion of Jews was not effective.[35] What Jews really needed to convince them was a miracle, which Gregory was able to provide later in the case of the conversion of a Roman. Even miracles visited on Jews seemed intended to persuade Christians more than to truly convince Jews. For example, when a Jew challenged the efficacy of dirt from the tomb of Saint Martin to heal a fever, Gregory asserts that "the Jew was afflicted with this illness and disturbed for a year; yet his wicked mind was never able to be converted through these torments."[36] In the absence of miraculous conversions, Gregory was resigned to the presence of the Jews as part of Merovingian society. That resignation likely reflects the stability of relations between Jews and Christians.

Gregory gives us glimpses of other Jews interacting with Christians and how difficult it is to characterize the Jewish experience in Merovingian society. In 584, Gregory tells us that a Jew, Armentarius, "accompanied by a man of his own religion and two Christians came to Tours to collect payment on some bonds which had been given to him on the security of public taxes by Injuriosus, who had been Vice-Count at the time, and by Eunomius, who had then been Count."[37] Armentarius can travel freely, although his two Christian companions may have been bodyguards or business partners. The Jew and the military leaders of Tours obviously had long and sophisticated financial dealings. Their personal relations seem casual and routine. They invited him to their dwelling, promising payment and gifts. Armentarius ate there with Injuriosus and at night moved to the other house.[38] It is not surprising to Gregory that Armentarius would share a meal with the two Christian leaders, suggesting that limits on the sociability between Jews and Christians were quite fluid. Unfortunately, something went very wrong, for Armentarius and his Christian companions were quickly killed by servants of Injuriosus.[39] Armentarius's relatives demanded that the case be heard in King Childebert's court. Ultimately, the relatives never appeared to challenge Injuriosus in court.[40]

The Jew's death is only part of the story's significance. We should not dismiss too lightly the ability of his family to get the case brought before Childebert. The ability to gain access to the king suggests real influence. That Injuriosus seems to have escaped is perhaps less important for our purposes. The relatives might have thought better of pressing the case or were stymied by the lack of evidence. In any event, they are not portrayed as powerless or abused. Furthermore, there does not seem to have been any particular moral to the account imputing special greed or deviousness to the Jewish financier. (We have to restrain ourselves from imagining that Christians of the day had internalized what are now considered to be traditionally derogatory images of Jews and money.) The whole episode is reported in Gregory's understated and straightforward style. If anything, it becomes another example of Gregory's obsession with the lawlessness and mystery of Frankish society. The Jew is an actor in the drama of Merovingian society not because he brings special meaning to the events but because he existed.

Jews had also learned how to exact vengeance in the best Frankish style. A quarrel apparently broke out between Priscus and "Phatyr, one of the converted Jews, who was godson to the King in that Chilperic sponsored him at his baptism."[41] According to Gregory's account, Phatyr attacked Priscus on his way to synagogue:

> One Jewish Sabbath Priscus was on his way to the synagogue, with his head bound in a kerchief and carrying no weapon in his hand, for he was about to pray according to the Mosaic law. (Perhaps Gregory noticed the kerchief because they only wore them on the Sabbath. His remark about carrying no weapon also suggests that Jews were usually armed.) Phatyr appeared from nowhere, drew his sword, cut Priscus' throat and killed his companions.[42]

Phatyr and his own companions took refuge in the church of St. Julian. Word was sent that the king was prepared to spare Phatyr (again the royal connections) but the others were to be killed. (It is not clear if these companions were other converts). Some kind of panic broke out among the group besieged in the church: "One of them drew his sword. Phatyr immediately fled, but the man killed all his associates and rushed out of the church, with his weapon still in his hand. The mob fell upon him and he was cruelly done to death."[43] Phatyr survived, and "received permission to return to Guntram's kingdom, whence he had come. A few days later he was killed in his turn by some of the relations of Priscus."[44] The Jewish ties of kinship and the desire for revenge mirrored those of the Franks. Again, it does not seem as if Gregory is censuring their actions because they were Jewish. Gregory, resigned to the state of his violent world, saw them as participants in the low-level conflicts that bedeviled Frankish society.[45]

On a larger scale, Jewish communities were resilient even when faced with radicalized Christian pressure. Local bishops sometimes targeted Jewish communities for conversion. In the famous episode of Bishop Avitus of Clermont's (c.572–c.594) conversion of local Jews, we can see that even these moments of conflict revealed that Jews were not simply the victims of Christian animus. The crisis was sparked by an attack on a convert to Christianity by some of his former coreligionists. Significantly, Jews felt secure enough to strike out at someone they perceived to be a traitor.[46] Even as a Christian mob besieged a Jewish synagogue, the language that Avitus employed, at least in Gregory's account, did not demonize Clermont's Jews but rather offered them membership in a unified city under his protection:

> On another occasion the Bishop sent this message to the Jews: "I do not use force nor do I compel you to confess the Son of God. I merely preach to you and I offer to your hearts the salt of knowledge. I am the shepherd set to watch over the sheep of the Lord. It was of you that the true Shepherd, who suffered for us, said that He had other sheep, which are not of His fold, but which he must bring, so that there might be one flock, and one shepherd. If you are prepared to believe what I believe, then become one flock, with me as your shepherd. If not, then leave this place."[47]

To be sure, there was intimidation, and it was effective. Some Jews, after much argument and hesitation, expressed their desire to convert. Avitus was able to celebrate a dramatic baptism. While the conversion of these Jews was presented by Gregory as a quasi-miracle that attested to Avitus's holiness and the new Christian nature of Clermont, his narrative emphasized the deliberative nature of the Jews' decision. Gregory takes their decision to convert as sincere. He can imagine them fitting easily into the fabric of local life with their new religious identity.

Gregory concludes his account of the Clermont episode with the comment that "those who refused to accept baptism left the city and made their way to Marseille."[48] This again emphasizes the importance of locality in the lives of early medieval Jews and the diverse experiences across the geography of Merovingian Gaul. The explosion of extremism by Avitus and some of his followers did not mean that every community in the region was marked by the same desire to efface a visible Jewish community. Jews could see that there were choices and those who still felt tied to their Jewish identity left for Marseille. As we saw in Minorca, there were places that remained open to Jews even as others began to see themselves—or wished to see themselves—as entirely Christian societies.

The local nature of the Jewish experience was what was important for Jews, despite the legislation directed at Jews from the various church councils of the Merovingian period. While the decrees regarding Jews—an extremely small percentage of the total number of laws issued—suggest a highly regulated theological state, it is more likely that the council decrees had no real observable force in social relations between Jews and Christians. It is important to resist preconceived notions of what these councils were. The very term *church council* suggests an ordered, systematic meeting whose rulings would be enforced by pious kings. The reality of the councils was more likely an ad hoc series of meetings with a changing assemblage of clerics limited to small geographic areas. It is not even clear if any of the council decrees were actually enforced or could have been enforced given the absence of a regular bureaucracy and police force. The laws themselves may have been issued in response to personal requests by local figures rather than as expressions of national or kingdomwide legislation.

At the most, the councils illuminate the perceived points of contact between Jews and Christians and suggest how some churchmen thought Jews and Christians should properly conduct themselves. The bishops seemed most concerned about regulating intermarriage, barring socializing during feasts and meals, and preventing Jews from owning Christian slaves. Jewish ownership of slaves provoked one council in 581–83 to declare that "complaints are being heard even now that certain Jews, inhabiting cities and towns, have grown so insolent and bold that Christians cannot be freed from slavery to them (though they demand it) even with money."[49] All of these points of putative scandal indicate, of course, how deeply interconnected were the two populations. The fact that Gregory of Tours ignores these texts strongly supports the impression that the councils were not the final arbiters of relations between Christians and Jews.

The integration of Jews into Merovingian society may have contributed as well to the evolution of syncretistic religious identities among Jews. Christians and Jews were thrown together in moments of great crisis. The threat of attack on a town by raiders or invaders often united cities made up of different religious and ethnic groups. Jews, for example, could be found defending the walls of Naples as an expression of their loyalty to the city's Ostrogothic rulers.[50] Just as there was not one stable Christian religious identity, there may not have been one fixed type of Jew in this early medieval world.

The conditions of the society that we have seen in Gregory's account suggest that there were many pressures on Jewish identity. The existence of converts would have pulled individuals and families to one side or the other of the confessional divide. Even temporary or expedient conversions may have left traces of Christian habits or outlook among Jews. The pressures of travel and interaction with gentiles likely created many situations of ritual compromise, if "compromise" was how Jews conceived of the practices. The entire society offered Jews examples of racially or ethnically diverse and fluid combinations of Roman and Frankish identities.

Did Jews of Gregory's day think of themselves exclusively as Jews wandering in a Christian world? Their names tell us very little. Were these Latinate names only assumed for dealing with Christians or did they reflect the internal naming choices of the Jewish community?[51] A seventh-century tomb inscription used the Latin names of Jews. It was only the candelabra and the Hebrew formula "Peace unto Israel" that identified the three *domini* as Jews.[52] One recent assessment of ethnic identity in this period concluded that "the people that our texts aim to measure with ethnic criteria are often in liminal situations, passing from one community to the other or being part of several overlapping systems of identification."[53] If Jews partook of other aspects of early medieval culture, then it is likely they experienced this fluidity of or at least potential challenge to a normative ethnic identity as well.

Moreover, we hear no echoes in Gregory or the other fragmentary texts of a Jewish longing for the Holy Land or a sense of the imminent end of their exile. These emotions would have likely been expressed in liturgical settings, but some perception of Jewish longing for the Holy Land would have been noticed by Christians even if only to gloat over its frustration. Christians who lived or passed through Palestine recounted the tropelike image of old Jews coming to weep over the ruins of the destroyed Temple.[54] But there seems to have been no echo of this depiction of Jews in wider Christian literature.

In the end, Jews and Christians of early medieval Gaul labored, traded, fought, and died together. They shared basic conceptions of honor and participation in local political life. Even episodic violence by radicalized Christians could not uproot the majority of Jews from their homes or make them feel like strangers in what they often felt to be their own land.

⚂

Jews had the deepest roots in Italy and by the seventh century remained a visible element in many localities. Italy was deeply fragmented by this time.[55] There were essentially three zones of influence. The Lombards, a non-Roman ethnic group, had moved into territory previously controlled by the Ostrogoths, the first non-Roman people that ruled Italy after the passing of imperial authority to Constantinople. Rome and some of the surrounding territory was ruled more or less by the pope, and in the south (along with Ravenna) the Byzantines had managed to hold onto territory recaptured under Justinian's devastating campaign to reconquer Italy in the sixth century. Again, this fragmented political geography underlines how specific local conditions would be of paramount importance in setting the tone of Jewish-Christian interaction.

This was an even more complicated society for Jews than that of Merovingian Gaul. It was a society under extreme pressure from warfare, famine, and plague. The population declined dramatically, urban areas contracted, and peasants turned to local strongmen or institutions such as monasteries for protection. Paradoxically, these crises probably acted in a way to unify local societies. Differences in religion became less important in the face of outside threats and challenges. Interestingly, we have no record of attacks or massacres of Jews associated with outbreaks of the plague or other periods of social stress.

The nature of Christianity itself should remain a prominent concern in assessing the experience of Jews. The Lombards shifted quite quickly from Arianism to Catholicism, suggesting that religious affiliations were still very fluid. Christianity may have idiosyncratically penetrated into the rural worlds of early medieval Italy. When Gregory the Great (ca. 540–604) tried to depict the heroism and sanctity of Benedict of Nursia (ca. 480–543), the founder of Western monasticism, he made it a point to include episodes of Benedict fighting in the sixth century against local pagans or impious Christians.[56]

Our foremost source for a fuller picture of Jewish life in early medieval Italy remains the letters of Gregory the Great, whose papacy gave direction to the early medieval church.[57] Gregory is often credited with codifying toleration for the Jews in church doctrine by supporting the rights of Jews to worship in designated synagogues and emphasizing their freedom from forced conversion. At the same time, Gregory insisted that Jews should not overreach themselves by having Christian slaves or by disturbing worship in churches. As with the church councils in Merovingian Gaul, these expressions of papal authority should be treated carefully. The range of papal power and influence was severely limited geographically and the ability of any early medieval government to enforce its policies was often haphazard at best.

Still, Gregory's writings give us hints of the complicated nature of relations between Christians and Jews and about the nature of Jews themselves, at least as they were seen by Christians. In an intriguing story in Gregory's *Dialogues*, he recounts how a Jew helped save a bishop in the days of the Goths. The Jew was traveling to Rome and took shelter in a surviving pagan temple: "One day, a Jew going from Campania to Rome was passing along the Appian Way. Evening was coming on when he arrived at Fondi. Unable to find lodging for the night, he decided to stay in the temple of Apollo which was close at hand. Fearing the unholiness of the place, he took the precaution of fortifying himself with the sign of the cross even though he did not have the faith."[58] It does not strike Gregory as strange that a Jew would be able to move freely about the countryside without any evident fear of attack. What bothered the Jew was the pagan nature of the temple. The Jew's use of the sign of the cross is perhaps a hint of his future conversion. Awakened during the night, the Jew witnesses, according to Gregory, a parade of evil spirits who have come to tell their master spirit what evil they have wrought. One of the spirits announces that he has stirred up Bishop Andrew to lust after one of his religious women living in his residence. The Jew makes his way to the bishop and forces him to confess his sins against the woman. Chastened, the bishop sends all his women away and dedicates a chapel to St. Andrew in the temple of Apollo.

The experience also brought the witness to salvation:

> The Jew, whose vision and rebuke had saved the good man, was in turn brought to eternal salvation, for, after receiving instructions in the mysteries of the faith, he was cleansed by the waters of baptism and brought into the fold of the Church. In saving his neighbor this son of Abraham attained his own salvation. Through God's providence it so happened that the preservation of the one from sin became the occasion for the other's conversion.

For Gregory, a Jew could play an important role in salvation. He was open to spiritual guidance and could be seen as an instrument of God's power. Gregory's story may have been his fantasy of what Jews offered Christians. However, in his relations with Jews there is the same quality of ambivalence that viewed Jews with anxiety but accepted them as part of a Christian world.

We can make only the barest guesses about how many Jews there were in early medieval Italy, where they lived, and their economic occupations. The traditional picture of Jews as merchants should not be allowed to overwhelm other evidence. There must have been rural or at least village Jews. Some attracted Gregory's attention in a letter to Peter the subdeacon in Sicily in 592. Gregory instructed his correspondent: "Because many Jews are settled on the estates of the church, I want some part of the *pensio* to be remitted to those of them who should wish to become Christians, in order that while they are attracted by this favor, the others, too, shall come up with such a desire."[59] How many more groups of Jews in these rural settings went unnoticed?

Woven throughout Gregory's letters about Jews are indications of how profoundly local conditions shaped the contours of their lives. Although Gregory seemed to tolerate Jews on church estates, Jews could be subject to harassment by local clerics. Jews often came to seek Gregory's support, which in itself is powerful testimony to their familiarity with the operative hierarchies of Christian society. For example, Joseph the Jew reported to Gregory that the bishop of Terracina evicted them twice from places in the local fort where they celebrated their holidays.[60] Several Jews from Massilia urged Gregory to complain to Vergilius, archbishop of Arles, from 588–610, and Theodore, bishop of Massilia in Gaul, about attempts in that region to bring Jews to the baptismal font by force.[61] Jews of Rome were able to petition Gregory on behalf of the Jews of Palermo where, they claimed, Christians were infringing on the rights of Jews to their place of worship.[62] The bishop of Palermo apparently evicted Jews from their synagogues and attempted to turn them into churches.[63]

Local conditions allowed some Jews at least to experiment with established religious conventions. Writing to Libertinus, praetor of Sicily in May 593, Gregory responds to a report that a Jew was trying to set up an alternative cult. Gregory recounts: "For it is said that Nasas, one who is the wickedest of all the Jews, built an altar in the name of the blessed Helias in a temerity that should be punished, and beguiled in a sacrilegious seduction many Christians to adore there."[64] It is a pity we do not know more about this episode. It seems to be testimony to the potential for syncretism between the two religious cultures. Living in an environment rich with shrines to holy men and women, this Jew Nasas seems to have tried to turn a Jewish prophet into a Christian saint. Gregory seems to be particularly concerned that such a shrine was already attracting Christians.[65]

The close connections of Jews and Christians are clear in the evidence of marriages across religious boundaries as well as the recorded conversions. In one letter Gregory discusses the case of a Jewish woman who has converted to Christianity and married a Christian. Gregory is writing to protect the wife from harassment—whether from Jews or Christians is unclear:

> We have decided, therefore, to recommend to your Experience Cyriacus and Iohanna his wife, the bearers of the present, in order that you should not allow anyone to oppress or aggravate them contrary to justice. . . . For whereas the aforementioned woman is said to be molested for having converted from the Jewish to the Christian religion after the reception of engagement gifts, . . . What was adjudged should be entirely and wholly observed, so that she shall not be persecuted by litigation moved by wicked men for the reason that she is known to have chosen the better part.[66]

If we knew how these two people met, courted, and negotiated the conversion and marriage we would have a much better sense of the dynamics of Christian-Jewish interaction.

The integration of Jews into local societies apparently brought some individuals to Christianity without any external inducements. Such conversions could spark local displays of zeal and aggressive piety. Gregory, for example, instructs the bishop of Cagliari to undo the actions of one Peter, a recent convert from Judaism. In an attempt to bring other Jews to Christianity, Peter tried to "convert" a synagogue to a church. According to Gregory, "accompanied by some undisciplined people, he occupied it on the day following his baptism, that is Sunday of the same Easter festivity, in a grave scandal, and placed there an image of the Mother of our God and Lord, a venerable cross, and white cloak that he had on when he emerged from the font."[67] Apparently both the bishop and the civil authorities acted together to stop Peter. Any attempt to upset violently the status quo drew Gregory's attention. In this case, his influence seems to have prevailed over local passions.

Some Jews had grown close enough to Christianity that they petitioned the church for baptism. With great pleasure, Gregory writes,

we have learnt from the Lady Abbess of the Monastery of Saint Stephen, situated in the territory of Agrigentum, that many of the Jews want to convert to the Christian faith under the guidance of the divine grace but that it is necessary that someone should go there with our instruction. We instruct you, therefore, by the tenor of this authority, that you should put aside all excuses, go to that place and hasten to help their wish with your exhortations, with God's favor.[68]

In another case, Gregory acted quickly to support several converts:

> It is appropriate that those whom our Redeemer deigned to convert himself from the Jewish perdition should be succored by us in a reasonable moderation lest they suffer (God forbid) want of food. We order you, therefore, by the authority of this command, that you shall not defer to give to the formerly Hebrews, children of Iusta, namely, Iuliana, Redemptus, and Fortuna, beginning on the next thirteenth indiction, each year ... *solidi,* which you should know that you must absolutely include in your accounts.[69]

It is interesting to note that Gregory has to encourage the missionary effort, suggesting that it was not a local priority. We do not know why these particular Jews wished to convert, but the general movement of small freeholders seeking security under episcopal or monastic patronage might have been part of their motivation. The impact of these conversions on the Jewish population also remains obscure. Were they marginal incidents or do they reflect larger numbers of undocumented conversions?

Christianity may only have been one aspect of how personal identities were constructed or perceived. An equally important factor in individual identity in this early medieval society may have been the division between free and unfree populations. While the large-scale chattel slavery of the Roman period was probably waning, the vast majority of the population was still in some dependent status to the great and powerful.[70] To what degree did Christian baptism overcome this more pragmatic reality of power and subjection is an open question. An individual's status, that is, his freedom or lack of it, may have counted for more in binding people together than common religious loyalties. We may be romanticizing the Christian past by imagining a society of Christians in opposition to small groups of Jews who remained on the outside of a common religious life. The loyalties and benefits of free status may have often trumped the disadvantages associated with Judaism.

Whatever social advantages Jews derived from slave-holding, Gregory was preoccupied with the scandal of Jews owning Christian slaves. He wrote repeatedly to clerics and rulers that such a state of affairs should not be tolerated.[71] However, Gregory could not risk undermining the slave-holding system itself. In 599 he wrote to Fortunatus, bishop of Naples, "But we have learnt from Basilius the Hebrew, who came before us with other Jews, that this purchase [of slaves] is enjoined upon them by various judges of the state and that it happens that among the pagans Christians too are purchased at the same time." Gregory felt it necessary to resolve these cases so that these slave owners who allege that they purchased slaves against their will do not suffer from the loss of the slaves.[72] Gregory could not intervene so drastically that the basic principles of a slave-holding society were undermined. A report of Jewish slave-holding prompted Gregory to remind the bishop of Luni in 594 "that, following the most pious laws, no Jew should be permitted to keep a Christian slave under his ownership."[73] The solution, however, speaks volumes of the need for stability in the agricultural regime of the period. Gregory continues: "these people, however, who are in their possession, although they themselves are free by law, because they have adhered to their lands for a long time for their cultivation and owe their condition to the land, shall remain and cultivate the land as they used to, and they shall pay to the said men *pensiones* and fulfill all that the laws demand from *coloni* and *originarii*."[74] Gregory seems to be sanctioning the conversion of Christian slaves of Jews to dependent laborers who could still rightly work for Jews. He also warned others against trying to claim or move the peasants. The Jews were not to be stripped of their "property" lest it set a dangerous precedent for other masters.

Jews of Italy thus faced many challenges and opportunities. They could look back on centuries of residence in Roman towns and territories that had traditions of protecting their rights to worship. Papal authority seemed largely committed to upholding these traditions even as some local Christian authorities tried to marginalize them. Jews seem to have escaped being reduced to slavery or abject dependence, although we have no clear sense of what kind of dependent relations they had with local patrons or authority figures. The increasing growth of the church institutions and Christian habits may have taken their toll on some Jews who sought refuge in conversion and the patronage of local bishops. The pressures of war, famine, and plague as well as tensions between the various political powers in Italy no doubt created periodic crises in the lives of Jews as well as Christians. Nevertheless, the few records we have suggest that Jews were finding their way among the many obstacles of this early medieval world. Gregory may have thought the end was near and that the ingathering of the gentiles was urgent. Jews seemed to have taken a longer view.

⚭

When compared to Minorca, Merovingian Gaul, and Italy, Visigothic Spain presents particular problems to scholars trying to reconstruct the nature of Jewish life and interactions with Christians. The traditional narrative of Jews in early medieval Spain describes a relatively open society under the Visigoths, an invading ethnic group that had adopted Arian Christianity. Conditions for Jews then took an apparently radical turn to persecution with the conversion of the Visigothic ruler to Catholicism in 589. The conversion of Reccared remains a vexing problem, but historians generally attribute it to Reccared's desire to form a kingdom unified by one faith that would bring together Visigoths and native Catholic Romans.[75]

There are several problems with this model. First, it remains as difficult for early medieval Spain as it is for early medieval Gaul or Italy to make any firm judgments about the degree or intensity of Christian culture beyond a small clerical elite that has left us written records. We risk distorting a very complicated social reality by assuming that a coherent Christian identity extended throughout the whole population. The ease of the conversion from Arianism to Catholicism suggests that religious identities were quite malleable. We have suggestive evidence that bishops in Spain continued to confront a vibrant pagan subculture well into the seventh century.[76] Given these factors, Visigothic Spain was likely not a monolithic society of devout Catholic converts with only Jews remaining as outsiders and potential targets of persecution.

The second problem is the enduring but misleading impression of Spain under the Visigoths as an efficiently ruled medieval polity. The records of Visigothic laws suggest that the anti-Jewish measures expressed society's consensus about Jews. The reality may have been more complicated. Visigothic Spain likely endured the same structural problems of any early medieval polity: the lack of established methods of communication; an ad hoc bureaucratic structure; no police force or standing army; and the problems of geography, corruption, and resistance on the part of the aristocracy to support royal policies. Regional conflicts and animosity remained even after the conversion to Catholicism.

The political history of the kingdom suggests that the kings were barely in control of their territory. After Liuva II succeeded his father Reccared in 601, he ruled for only two years before a conspiracy displaced him. For the rest of the seventh century the monarchy was elective and thus open to violent competition and dissension. Ten out of seventeen Visigothic kings in the century were deposed or murdered. One recent survey of the Visigothic kingdom has challenged

> the labelling of the Visigothic monarchy by some modern authors as an absolutist one. The label is totally unacceptable, however, not only because it has to do with a term which implies an anachronistic transposition of political ideas which would only appear much later, but also because socio-economic realities prevented the Visigothic monarchies from applying in practice the wide powers with which political-religious theory provided them.[77]

Against this background of political instability the anti-Jewish laws may not necessarily reflect the concentrated power of an early medieval state.

On the social level, there must have been relative freedom of movement and interaction between Jews and Christians. One saint's life records that the saint is noted for having built a *xenodocium* where doctors treated slaves, free persons, Jews, and Christians. The saint instructed that "the doctors should go through the entire city without ceasing and whosoever they found that was sick, be they slave or free, Christian or Jew, they were to carry in their arms to the *xenodocium*."[78] The Jews likely lived in a variety of locations, including rural settings. A recent scholar of Visigothic society has cautioned against seeing Jews exclusively as town dwellers: "But in the laws we catch glimpses of Jews of another and more humble sort; sometimes found living in remote areas of the country, they appear personally cultivating the fields and working as spinners and weavers, serving as slaves for other Jews and acting as the stewards and clients of Christians."[79] Ian Wood has also argued that power in Visigothic society shifted to patronage rather than only kin group loyalties, a change that would have given Jews more access to powerful people.[80]

With these complicating factors in mind, it is easier to imagine that the anti-Jewish laws of the councils and royal regulations present a misleading picture of an aggressive and effective theological state. The nature of the laws themselves, with their oscillation between persecution and tolerance, suggest a very limited impact on Jewish life. Even at the beginning, the scope of the Visigothic laws regarding Jews was quite narrow. Reccared limited his attention to confirming the canons of the Third Council of Toledo that called only for the forced baptism of children born of marriages between Jews and Christians.[81] There were no further legislative moves against Jews during the reigns of Liuva II, Witteric, or Gundemar (610–12).[82] In 613 Sisebut confirmed the baptism of children of mixed marriages and ordered that all Jews must leave his kingdom or become Christian.[83]

It is this kind of sweeping legislative decision that is so problematic to assess. We know that Isidore of Seville, the great scholarly bishop and no great friend of the Jews, opposed such a policy of forced conversion, suggesting that the support of the church for such an action was lukewarm.[84] What we do not know, of course, is how many Jews converted and how many fled. The fact that Sisebut issued the law without the support from a church council suggests that it was not a unified move that spoke for the cultural elite of the kingdom. It may have been a short-term effort to claim Jewish property by the king. One sign that the actions did not reflect a pan-Visigothic consensus was the opposition of Count Froga in Toledo to attempted forced baptisms by Bishop Aurasius, which may have been unrelated to Sisebut's decree.[85] It is hard to imagine that all Jews in the kingdom either converted or chose exile. Jews within reach of Sisebut's own men might have faced the choice but given the reluctance of lords to respond to the king's command to send men for his army, it is unlikely they were quick to fulfill his orders regarding Jews. Moreover, we know that Suinthila in 621 did not enforce Sisebut's anti-Jewish laws and allowed those Jews who had gone into exile to return. Converts could return to Judaism and Jews took up posts in the royal administration.[86]

Jews grew used to sporadic rhetoric that called for their persecution in the next two reigns of Visigothic kings. Sisenand's short and violent reign saw an attempt to purge royal offices of Jews. According to the laws enacted to enforce the purges, those Christians who refused to implement these policies—an interesting preemptive warning that suggests sympathy for Jews—were to be anathematized and excommunicated.[87] The equally unsettled rule of Chintila followed. He tried to purge royal government of Jews and to seize the property of Jewish slave owners. The animosity reached its peak in a ruling that again decreed that Jews must convert or leave Visigothic territory.[88] Again, since his rule lasted only three years, it is very difficult to assess the lasting success of these initiatives. Chidasuinth's attempts to enforce provisions against Jewish proselytizing argues that Jews remained in the kingdom and were considered a growing and active community.[89]

Reccasuinth, who ruled until 672, sought to constrain Jews already baptized to adhere to Christianity, which of course, suggests that there were converts who were returning to Judaism. He suppressed the *Breviarium Alaricianum* in 654, which had offered Jews legal protection to practice their religion. However, even when trying to undermine the legal standing of Jews, the king was aware that many elements in the upper echelons of Visigothic society would likely help Jews evade these restrictions. Again, he warned that priests or nobles or others who helped Jews in this way faced excommunication and confiscation of property.[90] The Ninth Council of Toledo in 655 had to reinforce the king's warning, suggesting that people were resisting the prohibitions against Jews.[91]

Even when a king's efforts against Jews were successful, the policy could be reversed. For example, Wamba (672–80) put down a revolt in Narbonne that had been supported by the local Jews and expelled them. However, he soon let them return. He did not seek vengeance against Jews in the laws of the Eleventh Council of Toledo.[92] King Erwig (680–87) returned to anti-Jewish rhetoric at the Twelfth Council where he sought to force all Jews to baptism and to mark out baptized Jews as inferior to other Christians. However, as with Reccasuinth's earlier legislation, the king had to devote an enormous amount of attention to potential resistance to these actions.[93] King Egica tried to rupture commercial relations between Jews and Christians by ordering Jews to sell to the government any property ever acquired from Christians.[94] The structures of early medieval government were ill suited to such an ambitious scheme.

Even more extreme, and equally unenforceable, was a decree before the Seventeenth Council in 694 accusing Jews of plotting with their coreligionists in Africa to deliver Spain to the Moors. All Jews except a few in strategic locations in the north were to be enslaved.[95] It is unlikely at this late date that the Visigoths could focus any sustained energy on the Jews. In 711 the struggles of the Visigothic monarchs came to a quick end with the lighting conquest of Spain by Muslims.[96] As we know, Jews survived to become a vibrant part of Islamic society in medieval Spain.

❈

Jews confronted many different types of Christian societies and many different types of Christians in these early centuries of medieval Europe. It would be unfair to the complex nature of their experiences to try to fit their interactions with Christians into a rigid thematic framework. Minorca, Gaul, Italy, and Spain all offered challenges and opportunities. Do we describe this experience as one of exceptional tolerance? How did the Jews themselves feel about their lives? Did they feel besieged by constant threats of danger as if they were living under the ever present shadow of Christian intimidation? Or were their responses and self-awareness dependent on the vagaries of changing situations? The common denominator perhaps is a certain resilience and fluidity in how Jews confronted an idiosyncratic Christian antagonism. The seeming confidence with which they negotiated these complicated societies must have derived from their sense of place and security. They did not behave as alienated and transient figures, living in these local worlds on the sufferance of Christian authority. They understood the levers of power in their local worlds, and they knew when they had to seek refuge elsewhere. In trying to understand the experiences of Jews and Christians in these micro-Christendoms, we should remember that the societies of early medieval Europe were complicated and ever-changing. Christians and Jews together moved through time in the early Middle Ages. For all of their variety, these centuries gave Jews a deep sense of their own localized "European" identity on the eve of the Carolingian age.

⚙ 2 ⚙

From the Carolingians to the
Twelfth Century

As CAROLINGIAN RULE became a reality, Jews had lived in Europe for hundreds of years and had survived scattered outbursts of Christian zeal. The domination of the Carolingians would not fundamentally disrupt the localized nature of the Jewish experience or the integration of Jews into European society. However, it would shape that experience in several interesting ways as the Carolingians created a more unified religious and political society. Carolingian dominance grew slowly as the family of Pippin and Charles Martel, beginning as the mayors of the palace for the last weak Merovingian kings, extended its rule over Gaul, northern Spain, Saxon territories of Germany, and northern Italy as the designated defenders of the pope. By the time of Charlemagne's crowning as emperor in 800 in Rome, the Carolingian empire was the most powerful political unit in Europe.

The rest of Charlemagne's rule was spent trying to keep this "empire" together as a coherent collection of territories whose inhabitants and rulers would acknowledge his overlordship. The basic dynamic of Charlemagne's strategy of government suggests that the stability of his reign was dearly purchased and perhaps a fragile reality. Constant campaigns to subdue and convert the Saxon pagans, raids on the Lombards in Italy, and defensive struggles with Muslims in the south and Vikings in the north, allowed Carolingian rulers to secure the loyalty of their military followers with gifts of plunder and land while holding their enemies at bay.[1]

Despite the seasonal violence of military campaigning, the impression that we have of Europe under Carolingian rule is one of increasing order, centralized administration, and religious reform. The self-conscious purification and regulation of Christian life, a central goal of Charlemagne's policies, emphasized establishing a common liturgy and monastic rule. This was vital to the Carolingians' vision of themselves as rulers of a pious Christian kingdom that merited God's favor. The cultural renaissance that flowed from this emphasis on correct liturgy and devotion produced new editions of the Bible, the fathers of the church, and works from classical antiquity. (Jews played important roles in the Carolingian revision of the biblical texts as "consultants" or teachers of Hebrew and in some cases as active scholars after their conversion.)[2] The networks of scholars at Charlemagne's court and in the various monastic schools linked the elite of the core lands of Carolingian Europe together in a way unknown since the waning days of the Roman Empire.

The idealized unity of the religious culture of Carolingian Europe did not seem to marginalize Jews. To be sure Carolingian exegetes and clerics produced a large literature drawing on patristic sources with anti-Jewish ideologies.[3] Still, many representations of Jews in Christian polemics were relatively mild.[4] Indeed, Jews were often overlooked by Carolingian Christianity. Contemporary Jews were seen as Pharisees and not as the true heirs of the Israelites. They thus offered no challenge to the assumption of the Israelite mantle by the Carolingians themselves.[5] For Carolingian propagandists, who imagined themselves and their rulers to be Israelites led by new Davids and Solomons, preserving an uncorrupted image of the Israelites

44

was crucial to their self-image. The cleric Druthmar, active in the middle decades of the ninth century, managed to delink contemporary Jews with the Israelites. He asserted that Jews were equivalent to Pharisees. In fact the name *Pharisee* meant the Jewish people.[6] The ancient divisions among Jews were replaced by a pharisaic unity that at the same time freed the original Israelites and their Christian heirs from any taint of Jewish or pharisaic corruption.

Jews themselves, even in their pharisaic incarnation, did not seem to challenge Carolingian Christian identity. The Christianity of the elite was focused not so much on the historical moment of Christ's crucifixion as on maintaining the correct standard of Christian behavior to guarantee God's favor. This meant that they were concerned with the proper liturgy, monastic rules, corrected texts of the Bible, education of monks and clergy, and the imposition of sexual norms. For the great mass of people in the countryside, it is safe to say that their contact with Christianity had less to do with the theological subtleties of resurrection and salvation than with the powers of local saints to offer healing and protection of crops, and the increasingly common experience of baptism.[7] These were all external observances. In this kind of Christian world, Jews were not central actors in the drama of personal salvation. The central struggle of Carolingian religion remained the implantation of Christian institutions and habits among the people of the countryside as well as the effort to convert—at least in name— the mass of pagan peoples on the borders of the empire.

It is difficult to know what impact the hierarchical, militarized ethos of the Carolingians had on Jews. We should not be hasty in imagining Jews as excluded from this militarized world. Given what we have seen of the ability of Jews to defend themselves in the rough-and-tumble world of Merovingian society and their future role in defending German towns, it makes sense that some Jews adopted the warrior habits of ninth-century Carolingians. We know that Jews held allodial lands (lands held freely) and could have served in the local levies.[8] It is too easy to project backward an image of the Jew as vulnerable moneylender isolated from the mainstream of Christian society. Daily survival required violence, and Jews needed to project an image of strength as much as anyone else not laboring as a dependent peasant. Their involvement with the slave trade—even in the more limited fashion that Michael Toch has documented—also suggests they had the ability to deploy violence when necessary.[9] This may be a distasteful echo from the past, but it reinforces the degree to which Jews were natural participants in the culture of early medieval Europe.

Other aspects of Carolingian culture reinforced the connection of Jews to local and regional societies. The position of Jews in these smaller worlds could often draw the anger of the churchmen. In the eighth century, Pope Stephen III wrote to Aribert, archbishop of Narbonne, that he was concerned about Frankish kings ceding allodial lands to Jews in southern France. These lands were intermingled with lands held by Christians both inside and outside of the towns. The pope feared the influence of Jews on the Christians who worked these lands or resided with them in the towns. He wrote: "We were, therefore, grieved and mortified when you informed us that the Jewish populace, ever rebellious against God and scornful of our rites, possesses hereditary allods in vills and in suburban estates—just like the Christian inhabitants—within the boundaries and territories of the Christians by reason of certain orders of the kings of the Franks." Jews, he reminded the archbishop, were "the enemies of God." Their proximity to Christians created threats: "there is danger in trading with him; and that Christian men cultivate their vineyards and fields, and that Christians, both males and females, living together with them—the treacherous—inside and outside cities are day and night defiled by words of blasphemy, and the miserable men and women perform uninterruptedly all imaginable services to the said dogs."[10] The pope paints a picture of social interaction between Jews and Christians that, in his mind, created a real risk to the salvation of Christians. At the same time, of course, such anxieties show how easy it was for contemporaries to imagine that relatively intimate relations could be established across the religious divide.

Jews had been part of the local towns and villages for generations and even the growing sense of a more unified Christendom was not enough to radicalize most Christians to turn against local people. The small numbers of Jews must have meant that most Christians never encountered a Jewish man or woman. Still, there must have been a general social consensus that accepted Jews because it is unlikely that any early medieval government would have been able to regulate relations between Jews and Christians if there had been unending antagonism.[11] Therefore, legislation protecting Jews had modest goals. The few surviving Carolingian privileges confirming the right of Jews to hold property and forbidding the conversion of their slaves without permission were granted only to individual Jews.[12]

When Jews appear in our records, they usually do so as merchants. The temptation, of course, is to view Jews as proto-urban merchants, a kind of preparation for the socially isolated moneylenders associated with the later Middle Ages.[13] Jews played a role in the continuing trade in luxury goods and slaves, which had a wide geographic range from eastern Europe to Venice. However, trade was still mostly the local or regional exchange of agricultural goods. Even if Jews were disproportionately active in some kinds of trade, their activities would have linked them to local networks as much as more traditional agricultural activities. Their role as merchant would not have made them a separate kind of human being removed from other European Christians. They would need patrons, servants, connections, and familiarity with local markets. In addition, we have already seen evidence of the varied occupations of Jews in previous centuries, and it is reasonable to conclude that that sociological variety of Jewish activities continued into the Carolingian period as well.

Moreover, when there were flare-ups between Jews and Christians, they seemed contained. The conversion of the deacon Bodo to Judaism (c. 839) may have scandalized some people but if there were any violent aftershocks they do not survive in the documentary evidence.[14] Even the author of the *Annals of St.-Bertin,* who seemed so outraged at Bodo's treachery, does not record any punishments or penalties inflicted on Jews for Bodo's actions. Other places where we might have expected retributions against Jews are interestingly quiet. For example, the St. Bertin annals relate that Jews betrayed Bordeaux to the Danes in 848 and Barcelona to the Moors in 852.[15] Neither account is accompanied by any indications of revenge attacks on remaining Jewish communities, which suggests that the accounts reflect a very limited anti-Jewish sentiment that may have inflated or falsified the episodes. In the same vein, the annals report that in 877 Charles the Bald was poisoned by his "Jewish doctor Zedechias, whom he loved and trusted all too much," without any account of vengeance or retribution. This suggests again that while Jews may have been convenient targets of animosity for some Christians (at least those who wrote annals) those feelings did not explode into wider violence.[16]

In the same way, Agobard of Lyons' (769–840) famous attacks on Jews in letters to Emperor Louis the Pious (778–840) also seem to have disturbed very few people.[17] Agobard's indictment of Jews may have been a way to vent frustration against the *magister judeorum* who controlled ecclesiastical property in

Lyons, and his vitriol against the Jews faded when it proved an unsuccessful tactic.[18] (His successor in Lyons, Amolo, had to restate complaints about local Jews who still employed Christian servants and about the attraction of rabbis as preachers.)[19]

Criticism of Jews never really gained any traction among the Carolingian political elite. The complaints themselves suggest a Jewish community—at least its elite members—with close ties to the imperial court. We catch glimpses of these Jews who made themselves useful to the king and court. Emperor Louis granted trading (and slaving) privileges to a few Jews who resided in Lyons; to Abraham, a Jew of Saragossa; as well as others who had no specific residence.[20] Charles the Bald (823–877), for example, sent silver to Barcelona for the repair of the cathedral by his Jewish *fidelis*, Judas.[21] Some of these merchants must have stimulated Louis's anxiety about proper behavior for he ordered the homes of the merchants, and those of his officials, prelates, and vassals, searched every week for male or female prostitutes.[22]

What we can see from this brief overview is the continuing attachment of Jews to their homes in Carolingian Europe. The borders of the Mediterranean remained open. As Einhard tells us in his annals, Isaac, a Jew, was sent on an embassy to Harun al Rashid, the caliph in Baghdad, probably as interpreter for the Frankish ambassadors. Isaac returned with gifts including the famous elephant.[23] Other Jews did not try to take advantage of this freedom of movement. We have no record of any mass migration of Jews from Christian lands to territory controlled by Islam.[24] Jews thrived in the new Islamic kingdoms of Spain, but clearly many communities of Jews chose to stay in Christian lands. This makes sense, for despite the fact that the center of gravity of Jewish population and culture remained in Islamic lands, the aggressive expansion of Islam may have frightened some Jews. It may have seemed better to take their chances in the less ideologically and politically coherent Christian world.

The choice to stay in Europe may have been motivated by other considerations as well, including, if Michael McCormick is correct, the attractive economic opportunities of trade between northern Europe and the southern Islamic lands. It was home, and Jews had lived in many of the communities for centuries. There could have been no realistic hopes of an end to exile given the great gains that Islam had made since the seventh century. At the same time, as in earlier periods in Merovingian Gaul, Jews were able to use the network of local communities as a way to survive any outburst of harassment. We know that at least one cleric in Lyons thought that the Jews of Lyons had smuggled some children out of that city to Arles when threatened with forced conversion, and that the same tactic was employed by Jews of Vienne and Chalon. We have no records of mass conversions, which surely would have been trumpeted by the church.

Jewish communities survived into the tenth century, and at least some Jews had the standing and power to antagonize competitors as far from the Carolingian heartlands as Venice. Peter II, the doge of Venice, wrote in 932 to King Henry I of Germany describing Jewish-Christian conflicts in Jerusalem with an appeal that Henry order the Jews in Germany to be baptized and if any refused, to remove them from participation in commercial activities. McCormick sees this as a Venetian attempt to quash continuing competition from Jewish slave traders and merchants.[25] (Although it is not clear whether the competition would have been eliminated if the newly converted Jews continued to trade.) In any case, well into the tenth century, there was an active group of Jewish merchants in northern Europe that generated anxiety all the way to Venice.

We should not see Jews of Carolingian Europe as a vulnerable and anxious minority. The knitting together of a Christian consciousness in Carolingian Europe probably had some effect on the self-awareness of Jews.[26] They could not have been oblivious to the more aggressive and unifying aspects of Carolingian government. What effect could such a growth of Christian culture have had on Jews? We have to remember that, given the variety of Jews who lived in the Carolingian world, there may not have been one Jewish response. Some Jews may have viewed the Carolingian ascendancy with alarm. It may have seemed like the growth of an aggressive Christianity that echoed the expansion of Islam. On the other hand, the sense of their own Judaism might have been strengthened as they saw Christians across Europe bound together by an increasingly common culture. The appreciation of the power of a unified religious culture may have prompted the establishment of recognized norms of rabbinic Judaism in the various European communities, which began in this period. Jews of Europe may have been thus particularly receptive to the growing enthusiasm for the Talmud and its ideal of a universal, normative Judaism. This sense of interconnectedness to a larger culture and religious world may have given Jews the resilience to resist conversion and assimilation into Christian society.

In many ways, the unity of Carolingian Europe was very brief, although it long held a central place in people's memories. The internal divisions that followed Charlemagne's death and the subsequent dissolution of the empire into local kingdoms set the stage for the next period in relations between Jews and Christians. At the same time that a meta-European culture was becoming more homogenous, the local polities and societies that made up Western Christianity were becoming more diverse. As the regions of the Carolingian empire coalesced into small territories and kingdoms and created local responses to incursions by Vikings in the north, Hungarians in the east, and Muslims in the south, Jews continued to cast their fate with their Christian neighbors. If Jews felt that they were now living in a world that was increasingly unstable and where most people were at the mercy of the powerful, then that was a feeling shared by Christians as well.

⊗

The history of European Jews during the century and a half after the dissolution of the Carolingian Empire remains obscure. This is not so surprising when we remember that understanding most aspects of this period is made arduous by the lack of evidence. This darkest of medieval periods is also one of the most crucial in terms of fundamental changes in European society. The seeming cohesion of European society under Carolingian rule quickly evaporated in the tenth and eleventh centuries. The societies that emerged by the eleventh century in the core countries of western Europe were radically different in terms of Christian identity, political authority, and economic structures.

Even if we assume that the degree of royal centralization under the Carolingians was exaggerated or romanticized, the polities that emerged around the turn of the millennium were societies where power was in the hands of local leaders. These "lords" exercised authority not as the regional representatives of kings but by right of their position as the military strongman of a certain land. The examples of the dynamic kingdoms of England as well as the increasing effectiveness of Ottonian rule in Germany show that traditions of royal government survived the demise of the Carolingians. The kings likely retained some of the power and cachet that had accrued to Charlemagne and his descendants. The experience of France, with its relatively weak early kings, should not be taken as the standard of European development. Nevertheless, real power in these societies devolved into the hands of lords who imposed judicial authority and economic control in their own names. In the ninth and tenth centuries these rulers bore the brunt of defending against Viking raids, Islamic incursions in the south, and Hungarian attacks in the East.

By the end of the eleventh century the basic structures of this new society of lordship were in place. The vast majority of rural people were now under the control of local lords. They were not slaves per se—for chattel slavery had declined over the centuries—but virtually all aspects of their lives were controlled by lords. Freeholders still remained but most of the core countries of Europe operated on a system of labor services, rents, and limitations on peasant freedom, inheritance, movement, and marriage. Scholars continue to argue whether mutation or revolution best describes the shift and when this occurred. Whatever the particular process, in the end, military leaders with dreams or memories of aristocratic ideals (and church institutions such as monasteries) accumulated enough influence to dominate local societies.[27]

Changes in Christian culture as opposed to local power relations were equally dramatic and reflected the same tension between a unifying Europeanwide culture and local variations. The tenth century is usually seen as a period of retrenchment for the institutional church as it struggled against the depredations of local lords and continuing pressure from non-Christian peoples. The institutions of the church survived this period of instability. Fueled by the surge in population and economic growth, after the turn of the millennium, new churches and monasteries were being constructed throughout Europe. At the same time that the church was growing physically, clerics and others had begun to think of the church and society in new ways in the eleventh century. This movement of reform was an effort to redefine the boundaries of church and secular society. By trying to purge the church of the interference and control of rulers and lords, most notably in the conflict over the investiture of church leaders, this movement created spheres of behavior that could be understood as sacred and secular. Although the traditional terminology of church versus state may be overdetermined, there was by the year 1100 a sense of society as divided into separate spheres or at least, contested spheres of influence. The attempt to free the church from external interference that marked reform created some social spheres for a secular identity outside the formal bounds of Christian religious life. It made European society in some sense more hospitable to Jews who did not have to be part of a universal religious society. These changes allowed Jews social space for their own identities even in an increasingly Christian world.

As with our earlier period, we have to make do with fragments of evidence in trying to assemble a coherent picture of how Jews lived in this period of social and cultural change. In terms of geography, the Jews still lived within the core countries of Europe. The large Jewish communities of Spain were now in the orbit of Islamic culture. We know of no Jews in the Scandinavian countries, and only small groups in the eastern kingdoms newly converted to Christianity. Their residence seemed restricted to the southern areas of France, Italy, and then later the Rhineland towns of Germany. By the end of the tenth century, there were probably four thousand to five thousand Jews. That number would grow to approximately twenty-five thousand before the First Crusade.[28]

Even within these relatively restricted parameters, by the eleventh century Jews were at home in Europe. They were increasingly acculturated to European society and rooted in different locations. As part of this acculturation, they shared a common cultural discourse with Christians. Scholars have long recognized this reality for Spain. Even in the most obscure period of the tenth and early eleventh centuries, we can get a sense of how Jews found niches for themselves in local societies that were continually evolving. The evidence of Jewish land-holding suggests a larger range of hidden relationships with Christians as servants or as fellow farmers sharing the risks of weather and market.[29] The violence and instability of this early medieval world shaped the lives of Jews as well as Christians. Rabbinic accounts of Jewish merchants who profited from the plunder of local conflicts suggest that some Jews were adept at exploiting opportunities in this dangerous world. Some Jews also engaged in the same kind of internecine violence that was tormenting the Christian community. However, even these rather tragic episodes speak to the interactions of Jews and Christians on very intimate levels: the acculturation of some Jews to medieval habits of violence and their ability to capital-ize, even temporarily, on such disorder.[30] Some Jews had grown so close to Christians that they left Judaism altogether.[31]

By the tenth century Jews were finding ways to make peace with their condition of seemingly permanent exile. Despite what appears to be idiosyncratic violence against several com-munities in Italy, Jewish cultural voices were seeking a reconcil-iation with the Christian societies in which they found them-selves.[32] The tenth-century *Yossipon*, a Jewish retelling and expansion of the account of Josephus's *Jewish War*, fits the Christian Roman Empire into the schema of the four empires in the Book of Daniel. However, the account of Jewish suffering and defeat at the hands of these empires, Babylonia, Persia, Greece, and now Rome was foretold long before the advent of Christ. Thus it was not the Jews' rejection of Jesus that deter-mined their exile but some previous crime for which God was punishing them. At the same time as the chronicle explained their suffering without reference to Christianity, it also held out assurances of their redemption. Jews could live under "Rome" with a certain degree of confidence and pragmatism. Jews remained part of God's plan—a plan that would transcend the transient victory of Christianity.[33]

When Jews reappear consistently in the documentary evidence, we see an immediate integration into local communities. Such a powerful attachment to their localities in the shadow of the Carolingian world is imbedded in the eleventh-century memories of Ahimaaz, who has left us his account of his family's arrival and history in southern Italy. On the periphery of Carolingian Europe to be sure, and responding to Byzantine and Arab culture as well, the memories of the Jews who settled in Oria still reflect the deep attachment to their homes that kept Jews in Europe for centuries.

The *Chronicle of Ahimaaz* is written as a psalm of praise to the Lord. It tells the story of Ahimaaz's family, preserved by God after the destruction of Jerusalem, from the time they settled in Oria:

> With praise, I will glorify Him that dwelleth in heaven; that, in His grace and justice, safely guided my ancestors who came forth with the exiles that were spared in Jerusalem, and delivered them from destruction, children and elders, young and old, for the sake of His great mercy and the merit of the fathers of old. At all times they were protected by the God of heaven; shield and buckler has He ever been to my forefathers, and so may He continue to be to their children to the last generation.

With a religious precision, Ahimaaz writes that "now with great care, I will set down in order the traditions of my fathers." He proceeds to tell the history of the Jews as it culminates in his family. He is filled with pride that these people "who were brought on a ship over the Pishon, the first river of Eden, with the captives that Titus took from the Holy City, crowned with beauty," found a new home: "They settled there and prospered through remarkable achievements; they grew in number and in strength and continued to thrive."[34]

Ahimaaz concentrates almost exclusively on the great men of learning and mystical power. His description of Rabbi Amittai is representative of the portraits of these scholar-heroes:

> Among their descendants there arose a man eminent in learning, a liturgical poet and scholar, master of the knowledge of God's law, distinguished for wisdom among his people. . . . And he had a number of amiable and worthy sons, intelligent and learned men, scholars and poets zealously teaching, worthy disciples, men of merit and renown, masters of secret lore, grasping and applying the deeper truth of scriptures; adepts in the mysteries, fathoming the veiled principles of *Hokmah* and *Binah* and of all abstruse learning; wise in the knowledge of the Book of *Jashar,* and familiar with the hidden meaning of the *Merkaba.*[35]

It was through such men and their skills that the community would be preserved. They established great schools of learning: "From that city he (Aaron) journeyed onward and went to Oria. There he found tents (of study), set up by the rivers, planted and thriving like trees by the waters, schools established, rooted like cedars growing at the side of flowing streams. There contending and flourishing in the pursuit of study, master in public discourse and of learned discussion of the Law, were the distinguished scholars, the genial brothers, my ancestors."[36] It was in European soil that the genius of his ancestors flourished: "Among them Aaron established his home. His wisdom streamed forth, his learning flourished there. He revealed great powers, and gave decisions of the law like those which were given when the Urim were in use, when the Sanhedrin held court, and the law of Sota was valid."[37]

57

Such a portrait is surely romanticized and should be used with caution, but we can still learn from the spirit of the account if not from its details. These miracle-working rabbis were remembered as the heroes of confrontations with the Byzantine emperor and invading Muslims. They cured the Byzantine emperor's daughter, foiled the plots of Saracen chiefs, and were even taken up as powerful viziers. What they do reflect is Ahimaaz's sense of belonging to the world of southern Italy. They lamented the exile, as Ahimaaz remembers, but they did their wonder-working largely on European soil.[38] Ahimaaz's Jews were part of the local population and survived transitions between Christian and Muslim rulers. They left their descendants a legacy of holiness and identity that would, they hoped, sustain them in the diaspora.

⚙

In the north of Europe, the records we have of Jews are more prosaic, without the overlay of magic and miracles in Ahimaaz's text. But in their own way they indicate an integration into local society equally profound. Individual Jews were able to forge close relationships with powerful Christians. In the chronicle by Thietmar of Merseburg (written 1012–18), a Jew named Calonimus was described as an intimate of the western emperor Otto II. According to Thietmar, after Otto's defeat by the Saracens in 982, Calonimus saved the emperor's life by giving him his horse:

> Along with Duke Otto and several others, the emperor fled to the sea where, in the distance, he spotted a ship of the type known as a *salandria*. He hurried out to it on a horse belonging to the Jew Calonimus but the ship's crew refused to take him in and continued on their way. Returning to the safety of the shore, he found the Jew still

standing there, anxiously awaiting the fate of his beloved lord. When the emperor saw that his enemies had also arrived on the scene, he sorrowfully asked this man: "What will now become of me?"

The emperor did find another ship to take him, but we hear no more of Calonimus.[39]

Larger groups of Jews were able to find stable and viable settlements in the uncertain years after the millennium. Their success seems the product of an expanding urban environment and the desire of local authorities to profit from the increasingly mercantile economy. The ability of Jews to live among Christians is also powerful evidence that Jews could manage or compartmentalize their own ambivalent feelings about Christians.[40] The evidence of the tenth and eleventh centuries suggests, as Johannes Heil has recently argued, that

> they would have been willing to face the problems (*halakhic* and others) that would arise and to settle, which meant to invest in the foundation of communities, in the formation of intellectual life, in the building of communal, social and physical structures of respectable dimensions. If feelings of distance and conflict vis-à-vis the majority were indeed as marked as some scholars today believe, such achievements would have been most unlikely.[41]

For example, Jews were often noticed as mourners at the funerals of local churchmen. When Archbishop Walthard of Magdeburg (1012) died and his body passed a village, "there was also a large crowd of Jews, to whom Walthard had been like a father."[42] Jews were also remembered as prominent mourners at the funerals of Adalberon (bishop of Metz (984–1005) and Annon of Cologne (1075).[43]

By the eleventh century several charters designed to attract Jewish residents were granted to Jews by German kings or bishops. Jews had been recognized by Henry IV as important residents of episcopal towns that provided him with allies. In 1074, for example, he granted Jews and others in Worms exemptions from certain taxes, giving them an incentive to support his rule.[44] By far the most extensive document that underlines the integration of Jews into eleventh-century urban life was the charter issued in 1084 by Bishop Rudiger of Speyer. He invited Jews to live in the town because he "wished to make a city out of the village of Speyer."[45]

The charter begins with a provision designed, it seems, to protect Jews from potential violence of local people: "Those Jews whom I have gathered I placed outside the neighborhood and residential areas of the other burghers. In order that they not be easily disrupted by the insolence of the mob, I have encircled them with a wall."[46] The burghers themselves do not seem to be thought of as a threat. However, the charter recognizes that there is another group of potentially violent local people. It is not clear from the charter whether this protection is afforded the Jews because they are Jews or because they would be known to trade and have money.

The rest of the charter provisions suggest that whatever the concern about potential violence, the actions and movements of the local Jews were not radically restricted. They had the right to buy and sell and exchange gold and silver "both within their residential area and, outside, beyond the gate down to the wharf and on the wharf itself. I have given them the same right throughout the entire city."[47] Unless we imagine that they were constantly under threat of physical violence—and there is no suggestion that they needed constant guards—the freedom to move about the city suggests they led relatively normal lives as part of their workday routines. They must have had regular contacts with a wide variety of Christians in the town. They could hire nurses and servants. They could sell Christians slaughtered meat that was ritually unfit for Jews. In the course of regulating their affairs, the "Jewish leader" responsible for resolving quarrels amongst the Jews or against them, would have had to deal with the bishop or his chamberlain as well as any Christians involved in the disputes. Jews as well as the Christians must have imagined they would be long-term residents of the city, for the bishop has "given them out of the land of the Church burial ground to be held in perpetuity."[48]

Jews had responsibilities as well as privileges. They were trusted to guard their section of the town wall: "They must discharge the responsibility of watch, guard, fortification only in their own area. The responsibility of guarding they may discharge along with their servants."[49] The privilege granted is that the Jews had to perform guard duty only on their own part of the wall. Still, Jews must have been armed and familiar enough with a kind of militia routine to be trusted to help protect the town. It was expected that they and their servants (including Christians) would be able to work together in these assignments. Most of all, they were thought of as local people with a stake in protecting the town.

Jews embraced their new town, at least according to one contemporary Jewish account, with enthusiasm: "At the outset, when we came to establish our residence in Speyer—may its foundations never falter!—it was a result of fire that broke out in the city of Mainz."[50] The Jews remembered a long and deeply held attachment to Mainz: "The city of Mainz was the city of our origin and the residence of our ancestors, the ancient and revered community, praised above all communities in the empire."[51] Something happened in Mainz, for the Hebrew account records an outbreak of violence by the burghers against Jews: "All the Jews' quarter and their street was burned, and we stood in great fear of the burghers."[52] As Jews had done at other times, they were able to find another town in which they could settle. God was still watching over them: "We then decided to set forth from there and to settle wherever we might find a fortified city. Perhaps the compassionate Lord might show compassion and the merciful One might exhibit mercy and the All-Helpful might help to sustain us, as in fact He does this very day."[53] Jews looked on the bishop of Speyer as their savior: "The bishop of Speyer greeted us warmly, sending his ministers and soldiers after us. He gave us a place in the city and expressed his intention to build about us a strong wall to protect us from our enemies, to afford us fortification. He pitied us as a man pities his son."[54] As attached as they remembered being to Mainz, Jews forged a new bond with the bishop and the town of Speyer.

Just as Christians at all social levels had learned to do, Jews seemed to have learned how to seek out patrons and protectors in eleventh-century society. Several of the leaders of the Speyer community sought out the protection of Henry IV, the emperor in 1090. In response, the emperor declared: "Let it be known that certain Jews, Judah b. Kalonymus, David b. Meshullam, Moses b. Yekutiel, and their associates, came before us at Speyer and requested that we take and hold them under our protection, along with their descendants and all those who seem to hope for security through them."[55] What is interesting here is how Jews seem to group themselves into a quasi-feudal hierarchy. The lowly people hope for security through the great men just as dependents look for or are forced to find security with their lords. The patronage of the emperor was a way for Jews to integrate themselves into the larger body politic of the empire.

Some of the provisions may have been the king's way to short circuit attempts by other lords to force Jews into relative dependence. ("Henceforth no one who is invested in our kingdom with any dignity or power, neither small nor great, neither free man nor serf, shall presume to attack or assail them on any illicit ground.")[56] Their property is protected not just from trespass but from larger attempts to dispossess them. ("Nor shall anyone dare to take from them any of their property, which they possess by hereditary right, whether in land or in houses or in gardens or in vineyards or in fields or in slaves or in other property both movable and immovable.")[57] Jews may have been seeking protection against the actions of lords who during the eleventh century tried to expand the extent of their local power. It is interesting to note as well that Jews clearly still had interests and property in agricultural holdings. Their primary residence seemed to be in Speyer, but their wealth did not come just from trade. In fact, their trade may have been the movement of agricultural goods rather than luxury items.

The association of Jews with royal authorities would continue during the course of the Middle Ages. This tie is often presented as proof of the weak position Jews held in medieval society. On the contrary, such connections with the powerful, including kings, should be taken as evidence of how Jews integrated into medieval society. Every individual was bound to more powerful people either by choice or intimidation in the eleventh century seigneurial regime. In many ways the provisions about Jews remind one of protections afforded monks and clergy. They had a specific and recognized place in these local societies. By the twelfth century, Jews had weathered, along with Christians, the profound changes from the period of Carolingian rule to the advent of lordship, and finally the slow recovery of local monarchies. In the next two chapters we can see how deeply integrated Jews would become in the culture and societies of these new European polities.

❊ 3 ❊

Cultural Integration in the
High Middle Ages

To FOLLOW JEWS FROM THE ELEVENTH into the twelfth century is to cross an invisible and arbitrary dividing line. The twelfth century—a period stretching from the middle of the eleventh to the middle of the thirteenth century—is usually considered the great age of expansion in the Middle Ages. All areas of European life saw growth and movement. The most important of these included an aristocratic diaspora and conquest, the spread of bishoprics, the expansion of towns and settlements, the rise of more centralized states and royal governments, and the multiplication of religious institutions and identities. The extension of aristocratic power helped to fuel the general agricultural expansion that accompanied the rise in population. New villages and towns were settled, forests were cleared, the populations of towns rose, and a mercantile economy touched greater numbers of people.

The expansion affected all areas of Europe. The Normans conquered England and Sicily, and pushed their way into Scotland and Ireland. Warrior elites from German lands fought over Prussian and Slavic lands near the Baltic Sea. The Christian kingdoms of the north in Spain continued to press southward, with a series of victories over the Muslims. By the

beginning of the thirteenth century, Christians had reached Toledo. Italy remained a politically mixed territory with the Normans in the south. The emperor controlled some territories
and the papacy others. Cities such as Venice and Genoa exerted
a growing influence as well. While many Christians must have
felt an increasing sense of the unity of Europe and Christendom, the diversity of European societies and populations
must also have reinforced the differences between people, even
if they were all Christians.[1]

The opening up of European society in the twelfth century
brought with it a self-conscious movement to reform monastic
and Christian practices. This process triggered the multiplication of new religious orders as people experimented with new
forms of identity and vocation. These developments suggest
inherent conflicts between competing visions of Christian
identity. Local variations in liturgy, custom, and practice indicate that Christians sought different ways to express their spirituality.[2] This flowering of Christian identities likely helped to
take some of the pressure off Jews as a vulnerable minority.
The Christian world was already steeped in notions of corpo-
rate or group identity.[3] Jews were one more group in this increasingly complex mosaic. Christian pluralism rarely went so
far as to include Judaism as a legitimate means of approach-
ing God, but the sheer variety of Christian models, both normative and heretical, helped to undermine the monolithic
quality of Christianity.

Whether or not Jews were "accepted" by Christian society,
Jewish culture of the medieval period demonstrated close par-
allels to Christian thinking and religious habits. These parallel
patterns or habits can be seen, for example, in Jewish legal exegesis. The reinvigoration of dialectical discussion among Jewish
legal commentators was a kind of scholastic commentary on

the Talmud, paralleling the growth of the schools and universities in Christian circles.[4] This reaction to inherited legal traditions was part of the common discourse that Jews and Christians shared. According to Ivan Marcus, "both cultures exhibit a dynamic of reaching back into earlier or ancient times for relatively neglected early sources and transforming the immediate cultural landscape by reviving and adapting ancient learning in a way that appears to be revolutionary but is actually conservative and revolutionary at the same time."[5]

Jews also internalized fundamental notions of martyrdom and religious extremism that were current in Christian culture. The Crusade chronicles, the accounts of the attacks on Rhineland Jewish communities during the First Crusade, suggest that Jews shared in a common "Christian" sensibility about martyrdom.[6] The Jewish accounts of the 1096 crusader attacks show that Jews were embracing the values of intentionality and individuality usually associated with twelfth-century Christian culture.[7] Jews of the Rhineland in the twelfth century remembered, or at least the chroniclers did, the communities of 1096 as representing an ideal in religious devotion and the model of pious service to God. The sense of individual spiritual purity that marked Christian theology and religious reform is present as well in the chronicles. While the question of influence may be intractable, it is clear that the two groups were thinking in the same religious language. These were people who felt supremely confident about the rightness of their faith. Crusaders and Jewish martyrs both asserted that the violence they created was fulfilling God's will.[8]

There may have been a more recurrent "dialogue" between Jewish and Christian religious culture, a kind of dance where Jewish prayer and ritual responded to their Christian counterparts. Marcus has explored how Jewish rituals around the education of children echoed and "answered" Christian habits of education and religious training.[9] Jews also tried to answer Christian miracles with their own: Jews of Rheims for example, offered to take out the scrolls of the Torah to help bring deliverance from a drought. A Jewish chronicler asserted that a group of rabbis brought rain to the Holy Land in 1210 by their prayers.[10] The polemical value of these efforts only makes sense if Jews were aware of the meanings and symbols of Christian culture and felt that they had to be answered. They deployed the same kind of discourse, only using Jewish images and symbols.

Acculturation happened, or was demonstrated, in other ways besides responses along these parallel lines of ideology. When Jews and Christians met, for example, ideas and images were exchanged, even unwillingly. Jews were key interlocutors for groups of Bible scholars in Paris in the twelfth century; they offered Christians instruction in Hebrew as well as Jewish interpretations of biblical texts. Despite the obvious obstacles to agreement and mutual acceptance, they could speak about the same text with the same kind of religious language.[11]

The disputations between Jewish rabbis and Christian clerics were also more formal venues where the two cultures took each other's measure. These performances, while stressful for the Jewish participants, perhaps did allow some real exchange of religious ideas even if both sides flatly rejected the other's positions. Jews participated in the same discourse about religious truths, even if they and Christians came to different conclusions. From the perspective of the Jewish participants as well, their engagement in disputations (albeit coerced) illustrates a basic ability to relate to Christians on theological issues. They had to function in the same cultural world of contested religious truths. Jewish responses also show resilience and self-confidence even under pressure. In the same vein, the anti-Christian polemic in Jewish literature, as well as the clever and subversive illustrations in some Jewish manuscripts, also suggests a level of social comfort and familiarity with the tools of Christian culture that permitted veiled attacks on Christianity.[12]

※

The variegated nature of medieval Christian society and the increasing proximity of Jewish and Christian cultural norms created situations where Jews could more easily embrace Christianity. The experience of conversion itself offers powerful evidence of the extent of Jewish acculturation, either real or imagined. Converts were living proof that the barriers between the two faiths could be crossed. Sometimes those transitions were difficult—as scholars including myself have asserted. Christians may have become incensed over the arrogance of some converts or their descendants, but the search for converts never stopped. The topoi of the host desecration stories, which contained such violent images about Jews, always included the conversion of one or more Jews.[13] Christians always felt themselves capable of accepting Jewish converts even if the reality was more complicated. Robert Stacey's study of conversion in thirteenth-century England emphasizes that English kings, particularly Henry III, followed an activist conversion policy with the institutional support of a halfway house for converts, the *domus conversorum*.[14] The experiences of converts was mixed; some assimilated with very few problems and others—probably the most marginal or indigent from an increasingly beleaguered Jewish community—created an insular world for themselves in the *domus*.[15]

In other places Jews came to Christianity from a range of social spheres. In Germany, for example, individual Jews were known in the elite circles around local archbishops. The Jew named Joshua was highly regarded by Bruno, archbishop of Trier (d. 1124), as a *medicus* even before he converted. (It is clear that other Christians remained suspicious of the new convert's profession of faith.)[16] Other converts came into contact with cosmopolitan members of the imperial court, and it was remembered that one Jewish woman converted under the patronage of Burchard, *vicedominus* of the Church of Strasbourg.[17] There is evidence as well of marriages between Christians and Jews on lower social levels. When these crossings of the religious divide occurred, they usually were not accompanied by scandal. Local records are silent, for example, about the liaison between Sophia, the daughter of a Jew, and a member of the lower nobility.[18] The scattered evidence of other conversions in Europe suggests a range of responses, but the primary impression is that while there may have been discomfort and anxiety over some of the conversions, the door to integration in Christian society was still largely open.[19]

How Christians understood the conversion of Jews in the twelfth century provides another window into the cultural proximity of Jews and Christians. Amidst the cultural, economic, and political developments that energized European society in the twelfth century, the human being—or rather a self-awareness of the individual—received new attention from medieval intellectuals. Scholars have given this movement of thought the contested shorthand label, "the discovery of the individual," although "discovery of the self" may be more accurate.[20] Despite reservations by many scholars about the specific terms of the discussion and even disagreements among participants in the debate, it does seem that many men and women in the twelfth century saw themselves more intensely as individuals in relation to each other, as well as in relation to God. They looked with obsessive interest into the workings of their interior selves, and they began to study and reflect systematically on their own minds, hearts, and souls.

It is against this background of intense individual spirituality that the conversion of Jews to Christianity becomes another way to understand how Christians made a kind of intellectual space for Jews in their culture.[21] In the early Middle Ages, even though individual converts were sometimes still identified as former Jews, there was little attention paid to the interior dynamics of a Jew's shifting identity. In particular, where large groups of Jews were concerned, they seemed to integrate fully into Christian society; there was no reference to a residual Jewish identity that might have suggested an incomplete internal conversion. And, too, there was no need for the conversion to be accompanied by a miracle that legitimated the conversion. When Christians of the twelfth century, more sensitive to the inner process of conversion, needed some external confirmation of the change in a Jew's identity they increasingly looked for miracles to sanction the change in individual identity.

Most of the accounts of conversion that have survived from the early Middle Ages tell the story of a bishop confronting a local Jewish community with the stark choice of conversion or expulsion. Clerics—at least on the local level—did not seem particularly concerned with the internal sensibilities of the Jews when they converted. All in all then, the conversions of Jews in the early Middle Ages reflect a landscape where changes in religious identity were not subjected to searching scrutiny. Nor was there a need for an external miracle to confirm the conversion. A laissez faire approach to conversion makes sense given the general pattern of ethnic and tribal conversions in early medieval Europe. Conversion was an ongoing phenomenon, and it would have been impossible and unwise to question the validity of each conversion.

After the millennium, conversion was no longer an ever present event in the core countries of European medieval society. But in the stories that became representative of the culture of the twelfth century, the most secure conversions were those accompanied by a miracle. These divine signals testified to the changes in the interior identity of Jews. They confirmed that the journey to God, which Christians knew to be hazardous and difficult, had divine guidance. When a pope, for example, wanted to attest to the sincerity of a certain convert—in the milieu of a society that constantly expressed doubt about the sincerity of converts—he asserted that the conversion had been stimulated by the miraculous appearance of hosts.[22]

Other Christians of this time—preoccupied with searching out the recesses of the interior person—responded to the conversions of Jews by looking for miraculous testimony for conversions. Guibert of Nogent recounts the forced baptism of a Jewish child seized in Rouen during the First Crusade: "Then he was led to the font, and after the ritual prayers were recited, they reached the point where a candle is lit and liquid wax is dropped into the water. There was one drop in particular that traced in the water such a perfect sign of the cross that no human hand could ever have managed to trace anything of the kind with such a tiny piece of matter."[23] Guibert was still suspicious even though the sign was verified by the boy's patron and "the incident was confirmed by a priest, and both of them swore several times in God's name that the story was true."[24] Guibert admits that "I would not have paid much attention to this matter if I had not witnessed this child's extraordinary progress."[25] In the context of the boy's religious career—he went on to become a monk—the meaning of the tiny wax cross was clear: "The appearance of the cross at his baptism, then, was not a chance event but was divinely willed. It was a sign of the faith that would develop in this man of Jewish stock, a rare event in our time."[26] The miracle confirmed the interior change in the Jew's identity and helped to suppress Guibert's doubts.

Sometimes a miracle was unavailable and it made conversion a long and painful process. That seemed to be the case in the conversion of Herman-Judah, whose controversial twelfth-century autobiography is one of the most precious texts we have of the perceived experience of conversion.[27] Even without the miracle, Herman's account displays the obsessive concern with the interior life of the individual and its impact on Jewish identity. Herman, a young Jew in the Rhineland, prayed constantly for some divine sign that would support his growing attraction to Christianity. He embraced the offer of a holy man who had befriended him to show Herman a sign: "He proposed to me, with the greatest steadfastness, this bargain: that if, in proof of his faith, he sensed no burning while carrying (as is customarily done) a scorching iron in his bare hand, I would faithfully submit to the cure of holy baptism, the dark cover of all unbelief washed from my heart."[28]

Unfortunately, the bishop stopped the ceremony and rebuked the holy man for trying to solicit a sign from God. The bishop instructed them:

> You are never to ask or, above all, to yearn for some sign from God to promote this change. Certainly, it would be the easiest thing for his omnipotence to convert whomever he wished without any miracle, but only by the secret visitation of his grace. A sign which is displayed visibly to the external sight would be idle if he did not work invisibly through grace in the heart of a human being. And, indeed, we read that many have been converted without signs but also that countless others have stayed fast in infidelity after seeing miracles. Besides, he added, it has to be known that faith which is won over by miracles has either no or very little merit but that faith which is undertaken without any incitement of miracles, but with simple piety and pious

simplicity, has the most excellent merit before God and the highest praise.[29]

After countless hours spent listening to Christian clergy explain the Christological meaning of the Old Testament, after hours of fasting, prayer, and invoking the sign of the cross, Herman felt no closer to becoming a Christian. He was trapped in some kind of limbo, neither Jew nor Christian. There was no way to actualize or see the "action of God's grace in the heart of a human being." In the end, only the intercession of the prayers of two holy women—representing for Herman the church itself—somehow gave him the confident awareness of his faith. "Not at all much later," he wrote, "by their merits and prayers, so great a brightness of Christian faith suddenly shone in my heart that it entirely put to flight from it the shadows of all former doubt and ignorance."[30]

Herman knew that his conversion was difficult to understand: "Look at me," he exclaimed, "neither the explanation given by many concerning the faith of Christ nor the disputation of great clerics could convert me to the faith of Christ, but the devout prayer of simple women did."[31] The intercession of the women seems to me a rather weak attempt to replace a miraculous sign. The workings of the inner person, his mind, soul, and heart were still hidden from the eyes of Herman's fellow Christians. His autobiography was, as Sander Gilman has written, a way to justify his new identity.[32] Such an effort was necessary because the same question remained—could a Jew really change his interior identity?

By the twelfth century, then, the landscape of conversion to Christianity had shifted. It was harder for a Jew—at least in many accounts—to be accepted as a true Christian, or to be accepted without miraculous confirmation. The very human quality of his or her interior self made it difficult for a Jew to become a Christian in the eyes of other Christians. This was not because Jews were perceived to be fundamentally different kinds of human beings or less than human.[33] Rather, it was precisely because they were considered to be human beings with the same kind of interior life as Christians. Just as Christians struggled to come closer to God, so too did Jews have to cross what was felt to be a virtually insurmountable distance to come to a fuller understanding of God. They made that journey as individual human beings.

I have been primarily concerned in this chapter with the intellectual and cultural integration of the Jews in medieval European society. Their familiarity and comfort with the cultural signs of Christian thought helped make Jews integral actors in their local communities as well. That does not mean that all boundaries between Jews and Christians were overcome. Medieval Europe was not an open, pluralistic society in a modern sense. However, I hope the next chapter will show the high degree of personal connections between Jews and Christians that existed in the still face-to-face worlds of medieval Europe. By recognizing these relations and the modus vivendi that they created, it will be easier to understand how Jewish communities survived in the face of violence and persecution.

❈ 4 ❈

Social Integration

THE VARIETY OF LOCAL CONDITIONS in which Jews found themselves make any generalizations about their lived experience very tenuous and artificial. How much a part of European society each Jew felt may have been based on personal factors or individual personality. Christian society was variegated, dynamic, and diverse. It accommodated Jews in many settings and conditions. We surely distort the historical experience of medieval Jews by focusing only on evidence of their alienation. Feelings may have changed during one's lifetime or at different points in the liturgical year. Feelings of integration might have been replaced by alienation after prayers and reflection in synagogues on Tisha B'av (the day of lamentation for the destruction of the Temples) or after surreptitiously watching Christian processions during Easter festivities. A long-term relationship with Christian partners or interlocutors may have softened the level of suspicion and animosity. The crucial point is that there was a spectrum of interactions and sentiments experienced by Jews as well as Christians.

We can recover some sense of that range by looking at how Jews lived in various communities in Germany, England, and France in the core centuries of the Middle Ages. It is important to emphasize that the kind of social interaction that we associate with *convivenica* in Spain and to a lesser degree in Italy was present in the countries known more for their persecution and expulsion of Jews. Jews in these more challenging environments still maintained relatively stable social relations with Christians even as they faced different forms of violence and the avarice of kings.

Germany offers the first example of this Jewish integration into local societies as Europe entered the crucial twelfth century. Ironically, the evidence comes from accounts steeped in violence against Jews. The attacks on Jews in the Rhineland towns during the First Crusade of 1096 remain a central concern to historians of the Jewish experience in medieval Europe. Most often, as discussed above, the attacks in the Rhineland have been understood as the outbreak of a long-simmering, but newly energized, anti-Judaism. The chronicles themselves, however, can help us see how Jews had become part of local societies before the outbreak of violence. In addition, the chronicles suggest as well that the integration of Jews was reestablished relatively quickly after the violence.

The chronicles constitute a problematic collection of texts. There has been an ongoing scholarly debate about their historicity, with some scholars arguing that they represent later "midrashic" meditations on persecution and cannot be trusted as evidence of what really happened in 1096. Others argue for their reliability even though there are multiple voices and editors involved in the compilation of the texts. I cannot adjudicate the debate here, but I am comfortable using the texts for some kinds of evidence. In particular, it seems to me unlikely that even authors writing to dramatize in a liturgical fashion the sufferings of Jews would have felt the need to invent examples of Christians protecting Jews or evidence of positive Jewish-Christian relations before the attacks. They must have been told these things when they collected their "evidence" for the accounts.

The authorship of the chronicles is also complicated. There may be at least five "discernible voices" in the group of texts associated with the attacks on the Jewish communities. The two earliest are the Mainz Anonymous and the "Trier unit" of the Solomon bar Simson Chronicle. The later texts are from the author of the "Cologne unit" of the Solomon bar Simson Chronicle, the "editor" of the Solomon bar Simson Chronicle, and the "editor" of the Eliezer bar Nathan Chronicle.[1]

If we do not ask too much of this evidence, we can still see that the texts suggest how deeply connected Jews were to their local societies. Many of the bishops and political leaders tried—albeit unsuccessfully—to protect local Jews.[2] More important, the chronicles recognize that the bishop and other political leaders were human beings whose motivations were understandable. Jews believed in the possibility of aid from the archbishop in Mainz enough to organize a major bribe: "They agreed on the counsel of redeeming their souls by spending their moneys and bribing the princes and officers and bishops and burghers. The leaders of the community, notable in the eyes of the archbishop, then rose and came to the archbishop and to his ministers and servants to speak with them."[3] The archbishop then offered to protect their money in his treasury and their families in his chamber "until these bands pass by. Thus will you be able to be saved from the crusaders."[4] When the chronicler returns to the archbishop's actions, he says the elders bribed him.

> The [Jewish] community came, when they bribed him, and begged him, so that he stayed with them in Mainz. He brought all the community into his inner chambers and said: "I have agreed to aid you. Likewise the *burgrave* has said that he wishes to remain here for your sakes, to assist you. You must therefore supply all our needs until

the crusaders pass through." The [Jewish] community agreed to do so. The two—the archbishop and the *burgrave*—agreed and said: "We shall either die with you or live with you." Then the community said: "Since those who are our neighbors and acquaintances have agreed to save us," they tried to bribe Emicho, the leader of the Crusaders.[5]

In hindsight, the chronicler suggests that the acceptance of the bribe was part of a Christian plot to destroy the Jewish community, but it must not have been obvious to those who were there—or they were so desperate they had to try it anyway. At the same time, the chronicler also can seem unsure about Christian intentions. He goes on to say that "the archbishop gathered his ministers and servants—exalted ministers, nobles—in order to assist us. For at the outset it was his desire to save us with all his strength. Indeed we gave him great bribes to this end, along with his ministers and servants, since they intended to save us. Ultimately all the bribery and all the diplomacy did not avail in protecting us 'on the day of wrath' from catastrophe."[6] The crusader band entered the city, helped by some burghers, and stormed the bishop's chambers where the Jews fought for their lives. The chronicle recounts that the archbishop's men and the archbishop fled from the church.[7]

The same kind of effort at aiding or trying to aid local Jews can be seen later, according to Solomon bar Simson's account, when the archbishop of Mainz sent a messenger to escort Rabbi Kalonymous out of the city:

He called to him and said: "Listen to me, Kalonymous. Behold the archbishop has sent me to you to learn whether you are still alive. He commanded me to save you and all those with you. Come out to me. Behold with him are three hundred warriors armed with swords and dressed in armor. Our persons are pledged for yours, even to death. If you do not believe me, then I shall take an oath. He is not in the city, for he went to the village of Rüdesheim. He sent us here to save the remnant of you that remains. He wishes to assist you." They did not believe until he took an oath. Then R. Kalonymous and his band went out to him.[8]

The Jews found refuge for a while with the archbishop in the village. As the chronicler explains: "the archbishop was exceedingly happy over R. Kalonymous, that he was still alive, and intended to save him and the men that were with him." But they must have been pursued, and the archbishop abandoned them again. Faced with pressure from crusaders and locals, the archbishop pressed the Jews to convert. Instead Kalonymous and the group chose suicide, and after Kalonymous dispatched his own son, the archbishop turned against him; he lost the sympathy of the local villagers as well. The chronicle continues with a detailed account of the last stand of these Jews against a combined group of crusaders and villagers. Many seemed to have fallen in combat, while others converted.[9]

Other clerics tried to help their Jews. The Mainz anonymous confirms that the bishop of Speyer sought to defend Jews:

When Bishop John heard, he came with a large force and helped the [Jewish] community wholeheartedly and brought them indoors and saved them from their [the crusaders' and burghers'] hands. He seized some of the burghers and "cut off their hands." He was a pious one among the nations. Indeed God brought about well-being and salvation through him. . . . Through him all those forcibly converted who remained "here and there" in the empire of Henry returned [to Judaism]. Through the emperor, Bishop John removed the remnant of the community of

Speyer to his fortified towns, and the Lord turned to them, for the sake of his great Name. The bishop hid them until the enemies of the Lord passed.[10]

The bishop of Trier—without local allies—seems to have tried to save Jews before his resistance collapsed in the face of local intimidation: "The bishop was very frightened because he was a stranger in the city, without a relative or acquaintance. He did not have the requisite strength to save them." He warned them that not even the emperor could save them.[11] In the end the bishop and his few men, faced with the crusader and possible burgher attack, abandoned the local Jews and gave them the choice of conversion or expulsion from the safety of his palace: " 'You cannot be saved—your God does not wish to save you now as he did in earlier days. Behold this large crowd that stands before the gateway of the palace.' "[12] All of these accounts are problematic, but they reflect a persistent memory among the Jewish chroniclers that local church authorities were seen, at least at the beginning of the crisis, as offering a real source of support and protection.

The chronicles tell us how deeply Jews were embedded in their local worlds. No matter how quickly the crusader bands moved across the territory, many Jews could have fled. But they calculated—incorrectly as it happens—that they could best find security by relying on local defenses or the protection of local authorities. Kalonymous, one of the leaders of Rhineland Jewry, was able to send a mission to Emperor Henry IV in Apulia calling for help in opposing the violence of the crusader bands.[13] The quick movement of the mission, and the emperor's positive if belated response, indicate that Jewish leaders could maneuver among the German elites and that trying to arrange for rescue and refuge was not thought to be a futile strategy.

We can see evidence as well of the deep social connections between some Jews and their Christian neighbors. Whatever the extent of martyrdom, it is clear from the chronicle accounts that Jews sought refuge among Christian townspeople with whom they must have had some kind of relationship. In Cologne, for example, the chronicler recounts that when the violence broke out at the beginning of Shavuot, Jews "fled to gentile acquaintances. They remained there for the two days of Shavuot."[14] One woman, Rebecca, was slain trying to reach her husband who "had already left his house and was in the house of his gentile acquaintance." The Jews who managed to reach the houses of friends or neighbors were lucky: "But the rest of the community was saved, remaining in the houses of their acquaintances, to which they had fled."[15]

The chronicler remembers that Jews were confident that violence would not come from their neighbors without the provocation of outsiders. As the text records about the violence in Trier, "for up to that point, the burghers had never intended to do any harm to the [Jewish] community, until those pseudo-saintly ones arrived."[16] In Mainz, some of the burghers of Mainz fought against the bands of crusaders: "Then all of them [crusaders] came with swords to destroy us. Some of the high-ranking burghers came and stood opposite and would not allow them to harm us. At that moment the crusaders stood united against the burghers and one side smote the other until they killed one of the crusaders."[17]

In Worms, for example, several Christians pleaded with a particularly well-known Jewish woman (perhaps a courtesan) to convert: "There was also a respected woman there, named Minna, hidden in a house underground, outside the city. All the men of the city gathered and said to her: 'Behold you are a "capable woman." Know and see that God does not wish to save you. . . .' They fell before her to the ground, for they did not wish to kill her. Her reputation was known widely, for all the notables of the city and the princes of the land were found in her circle."[18]

In addition to the bonds between Christians and Jews that offered the hope of protection, the chronicles also suggest that in some cases even closer ties between individuals had evolved. Some Christians had even been attracted to the Jewish community. We have a glimpse of a convert to Judaism. In Mainz, "there was a certain man named Jacob ben Sullam. He was not from a family of notables. Indeed his mother was not Jewish. He called out loudly to all standing near him: 'All the days of my life till now, you have despised me. Now I shall slaughter myself.' "[19] Other converts saw the attacks as an opportunity to demonstrate their loyalty to the community. In Xanten: "There was, in addition, a servant of the Lord who was a true convert. He asked Rabbi Moses, the high priest, and said to him: 'My lord, if I slaughter myself for the unity of his great Name, what will be my lot?' He said to him: 'You shall sit with us in our circle, for you shall be a true convert and sit with the rest of the saintly true converts in their circle. You shall be with our ancestor Abraham who was the first of the converts.' "[20] Perhaps the comments were meant as criticisms of how Jews had treated the converts in more peaceful days. The fact that there were converts is intriguing evidence that some Christians had drawn close to the Jewish community.

The language of the chronicles emphasizes how deeply attached Jews felt to their localities.[21] They thought of themselves as the holy congregations of Mainz or Speyer. Simson describes Mainz as the shield of all the other Jewish communities. And when Mainz was attacked, he laments that

"Gone from Zion are all that were her glory," namely Mainz. The sound of "the lords of the flock" ceased, along with the sound of "the valorous who repel attacks," who lead the many to "righteousness." "The glorious city, the citadel of joy," which had distributed untold sums to the poor. One could not write with "an iron stylus" on a whole book the multitude of good deeds that were done in it of yore. In one place [were found] Torah and power and wealth and honor and wisdom and humility and good deeds, taking innumerable precautions against transgression. But now their wisdom has been swallowed up and turned into destruction, like the children of Jerusalem in their destruction.[22]

The scholars and leaders of these communities attain a standing that equaled that of the Temple: "from the day that the Second Temple was destroyed there were none like them in Israel and after them there will be no more."[23] These were not communities longing for the end of the exile. Jewish communities in Germany may have constructed a particular historical memory of their past, emphasizing or embellishing the nature of martyrdom where true martyrs had been only a small minority of the victims of the Crusade attacks.[24] In any event, Jews quickly set about the process of rebuilding their lives.[25]

The general tenor of the twelfth and thirteenth centuries seems to have been positive for Jews in German-speaking lands. They became important elements in the urbanization of medieval Germany. Their communities expanded in towns where they had established settlements as well as in new foundations.[26] Even well into the thirteenth century Jews seemed to have a foothold in trade, minting, and financial advising to German rulers and princes.[27]

Indeed, Jews of the *hasidei ashkenaz*, the so-called pietist movement in the twelfth and thirteenth centuries, presented their world as one of "relatively peaceful co-existence between

Jews and Christians."[28] The rabbis who led this movement were concerned to make sure that Jews not be seduced—as some were—by Christian culture, including the appeal of reading vernacular romances.[29] Such concerns surely point to a world where Jews, while aware of their minority status, could still feel themselves to be part of a Christian society.

Thriving Jewish communities could be found in many of the episcopal cities of medieval German kingdoms.[30] The account of Herman's conversion that we saw in the previous chapter demonstrates that in the generation after the Crusade attacks, Jews moved freely in German towns and had close relations with Christians up and down the social scale. For example, Herman's family sent him to stay with the bishop of Münster to watch over his loan. He engaged Rupert of Deutz in an impromptu disputation that is markedly civil. Herman's family was increasingly concerned about the time that he spent with Christians. He thought they were beginning to doubt his loyalty: "I had consorted with Christians, in such an eager and familiar way that I could now be thought, not a Jew, but a Christian." Even on the day of his marriage, the intermixing of the communities was clear: "When the day of the marriage feast was at hand, many gathered together there, not only Jews but also my Christian friends."[31]

⌘

Such testimonies to personal relations between Jews and Christians are rare, but even in the more bureaucratic records of England and France, we can find suggestive evidence that Jews truly belonged to the local societies in these Christian kingdoms. In England, for example, there seems to be strong evidence that many Jews earned their livelihood in small towns and villages. Often just one or two Jewish families lived in these places.[32] This suggests, of course, that they were comfortable living and working among Christians even far removed from the centers of royal power and protection. The presence of small numbers of Jews in more or less isolated towns and villages indicates a high degree of confidence on the part of Jews in their safety among gentile neighbors. The physical structure of Jewish settlement was conducive to interaction with Christians as there were no closed or isolated Jewish settlements.[33] When pressed Jews even ventured into churches and monasteries as Leo of Worcestser did when he was arrested for forcible entry into the Hospital of Worcester to get his local partner to pay up.[34] Jews and Christians seemed to socialize as well. One local bishop was incensed that Christians had attended a Jewish wedding.[35]

The royal officials who administered relations with the Jewish communities were in daily contact with Jews.[36] Jews had established relationships with lenders at all levels of rural society, and the majority seemed to be from manors and villages in the hinterland of major towns that acted as administrative centers for recording Jewish loans.[37] In the middle of the thirteenth century, Jews even seemed to forge local partnerships with monasteries in buying up debts that resulted in the transfer of land to the monasteries. This practice, which threatened the middle and upper middle ranks of the gentry and barons, prompted much of the mid-century baronial violence against Jews.[38] The extent of these widespread credit relations with individual Christians must have meant long-standing contacts and knowledge of local society. The absence of generalized violence, or the limiting of that violence to gentry debtors whose land was threatened, suggests that Jews were not living in a constant state of fear of most of their rural clients. Or at least, they were no more at risk of violence and crime than a Christian who engaged in the same practices and traveled the same dangerous roads. Even after moments of crisis, Jews set about trying to contract new loans or reestablish information about loans that might have been lost in the unrest.[39]

Jews may not have become English—if there was such a thing as a universal Englishness in the Middle Ages—but they had achieved a recognizable and relatively stable place in medieval England. They managed to secure a similar place in French lands. The experience of Jews in France suggests a resilience and attachment to local society that resisted, for a time, the pressures of French royal policy. That policy, or series of ad hoc responses to extract revenue from the Jews, ultimately forced Jews out of French lands. It is a grim picture of unrelenting royal animus. However, there are moments when it is possible to imagine an alternate reality, or at least a temporary reality that must have allowed interactions between Jews and Christians. Whether these relations reflected true mutual tolerance or respect, the social peace that existed allowed Jews to conduct various kinds of business that made them the target of royal avarice and religious anxiety.

Jews continued to find ways to live and earn their livelihoods even in the face of repeated expulsions and uncertain royal policy in France. The domains of the French king must have still held attractions for Jews as they regularly returned to these lands after their expulsion. After Jews were expelled from the royal domain in 1182, for example, they must have felt confident enough about their place in French society to return in 1198.[40] The approximately two thousand Jews of Normandy also decided to cast their lot with the new Capetian rulers of the duchy. French treatment of the Normandy Jews allowed for the peaceful transition to Capetian rule and the continuation of relations between local Jews and Christians.

Even royal restrictions on Jewish economic activity could not completely undermine the financial health of Jews. After the monarchy imprisoned and ransomed wealthy Jews in

1209–10, at least some rich Jews had rebuilt their wealth as we see from the marriage of Moses of Sens.[41] Despite the increasing obsession of Louis IX in the thirteenth century about Jews and his attempts to restrict their activities, we should not ignore the continuing vitality of Jewish economic and social life—and the difficulty that the French royal government had in enforcing its own restrictions against Jews. We have already seen that Jews continued to interact with Christians in various economic and social roles. Jews did not flee Paris or Capetian lands. Although there may have been some loss to the community, Jews entered the kingdom from England and Toulouse.[42] Jews retained enough vigor to be the target of Capetian takings in the future.

There must have been wealth and vitality in some of the Jewish communities despite the continuing pressure. The bishop of Béziers built or authorized the construction of a new synagogue in 1278. The king at the time—Philip III—ordered it razed but clearly there was demographic and social impetus for the new construction that was only limited by the Capetians in the last instance. The later policies of the crown, including a renewal of usury prohibitions, and bans on new synagogues, loud chanting, and the study of the Talmud, clearly indicate a continuity of anti-Jewish policy. On the other hand, the targets of these laws suggest a continued vitality of Jewish social life that is hard to reconcile with a community brought to financial ruin.

In the process of the royal moves against Jews, we can see that powerful elements in French society did not follow royal policy willingly. Kings and royal minions had to plan economic seizures and expulsions in secret. They were worried about obstacles and opposition from other lords, the flight of Jews, control of popular violence, and seizure of books to secure information on outstanding loans to Christians. In Pamiers and other places, the process was difficult for royal officials to control, and the collusion of the lords with Jews delayed the completion of the taking.[43] The king and royal officials recognized that Jews had forged bonds with Christians in their local societies.

Even this quick survey of Jewish life in Germany, England, and France should suggest that Jews did not live lives solely of alienation, marginalized from Christians in small villages or towns. The actions of royal governments in England and France protected Jews from outbreaks of some kinds of violence, but when those governments decided to exploit Jewish wealth, there was little defense in the long run. However, Jews in these countries were deeply attached to and engaged with their local societies. The long years of residence in the small towns of England, France, and Germany had created strong social networks between Jews and Christians. They had daily connections through business and personal dealings. It would take the concentrated force of medieval monarchs to break those relationships.

⚘ 5 ⚘

Violence

VIOLENCE IS TRADITIONALLY perceived to be at the core of the Jewish experience in medieval Europe. The history of Jews on one level is the story of rising levels of anti-Jewish polemic, accusations of atrocities, physical attacks, and finally expulsion from much of western Europe as Christian toleration of Jews declined. It is not surprising that this is the dominant image of the medieval Jewish experience. The information about Jews that has survived in the documentary record has usually been about instances of persecution or expulsion. Peaceful coexistence often went unrecorded. Moreover, moments of antagonism and violence occurred with a seeming regularity that makes it easy to see them as having been causally connected. In any impressionistic account, violence against Jews appears as a fundamental feature of Christian culture and society.

However, we have been accustomed to lament any violence against Jews without pausing to consider what kinds of attacks were directed against them. To undertake such a task is not to minimize the suffering of individuals or to excuse the violence visited on them. But people react to different levels of violence in different ways. In other words, we have to be careful not to imagine that every attack on Jews was seen by Jews—or by Christians—as representing an inexorable and unchanging anti-Jewish animus. As I hope to show, many Jews in medieval

Europe became quite adept at assessing the long-term threats and consequences of Christian polemic and social violence.[1]

The nature of violence in medieval society writ large likely affected Jewish perceptions of what constituted fundamental threats to Jewish safety. We need to understand how Jews viewed generalized violence in medieval society and how they measured it against their own experience. The polemics, attacks, and expulsions directed against Jews may not—in their minds—have been such egregious violations of social norms when judged against the larger context of social disorder or vicious judicial violence. This contextual sensibility is crucial to interpreting the reactions of Jews to anti-Jewish persecution in the Middle Ages. We are right to remain horrified at the suffering many individual Jews endured, but they made decisions—when they could—about their safety and collective futures based on what they knew of their societies. It is by trying to recover their reactions to the violence and how they thought to protect themselves (if they could at all) that we can best understand their experience.

Violence or the fear of violence, in myriad forms, was a constant that both Jews and Christians had to confront.[2] One ever-present concern was crime. Any collection of court cases from English *dooms*, plea rolls, or royal French investigations illustrates a society that witnessed continual crime from theft, assault, to more extravagant violence. Moreover, the judicial records only give a glimpse of what must have been the true levels of criminal disorder. The cases recorded were only the ones that the relatively limited judicial systems were able to document. Most crimes—even in relatively well-governed kingdoms such as England—went unpunished because there was no police force to apprehend criminals.[3] Local pressure, reconciliation, lynch mobs, ad hoc prosecution, or extraordinary efforts by royal authorities were the only mechanisms to bring suspected criminals to justice.

Violent crimes also always carried with them the threat of revenge killings by a victim's kin or friends. The legal punishments for crimes were also characterized by a high level of violence. The dispensation of justice was often surrounded by a penumbra of dramatic cruelty and physical suffering.[4] Such criminality did not paralyze medieval societies—and we should be careful not to create a caricature of the violent Dark Ages—but recognizing the presence of real violence casts a different light on how people of that time perceived order and stability or their absence.

Another ever-present factor in the calculation of violence and stability in medieval society was the open and hidden pressure applied to the rural population by the lords and their henchmen.[5] Such intimidation and the visible violence associated with it varied from place to place and in different periods. Although the eleventh- and twelfth-century economic and demographic expansion prompted lords to offer increasingly attractive terms to groups of rustics or urban pioneers to settle new lands or towns, the threat of violence was always present. The image of a mounted lord and his retinue, descending from a local castle, could never have been far from the minds of most medieval people.

The low-level warfare between competing lords or between kings and their putative supporters was also a source of social disorder that could flare into open violence. Even if most medieval warfare was a limited affair until the *chevauchées* of the Hundred Years War, the movement of armed men through the countryside was always alarming. The trajectories of this violence often depended on the strength of royal governments, but even strong English and Capetian kings had to contend with potentially disruptive factions or groups of nobles. In German territories or in contested areas, Jews would have witnessed regular struggles between military forces. Jews had to make their way in these societies with all of this real and potential violence. Antagonism and threats directed specifically against them were only part of their daily calculation of personal safety. In the particularly turbulent years of the fourteenth century, Jews would have seen violence and social disorder affect large groups of Christians.

It was not just "background" violence that affected how Jews assessed their own position in medieval society, but the level of polemic and negative imagery current in medieval discourse and culture as well. It is easy for us to be shocked at the language and images that were used to describe Jews by many Christians. The language condemning them as deicides, murderers of children, cannibals, desecrators of the host, agents of the devil, and animals with distorted features makes Jews seem a universally despised group among Christians.[6] But we often see these images and language in isolation from the rest of medieval rhetoric and discourse. Jews were not alone in being described in terms of negative images. Christians were used to describing and denigrating each other with a vigor and venom that rivaled their descriptions of Jews.[7] Orthodox Christians used language to indict heretics that often overshadowed their rhetoric about Jews. By the end of the twelfth century, as one historian has written, "Christian heretics attracted so much attention and appeared so much more urgent a problem, that the Jews seemed to have diminished in importance to Christian apologists, and we find Alan [of Lille] tucking them in after the Cathars and Waldensians almost as an afterthought."[8] Christians had many "devils" to choose from in addition to heretics. The Mongols in the thirteenth century traumatized much of eastern Europe. Muslims were a constant threat even when Christianity was aggressively confronting them in crusading efforts. All of these groups were regularly described in language comparable to that used to defame and demonize Jews.[9] Many disputes between popes and emperors, different kinds of monks, and lords and dependents were characterized by intensely insulting and vicious language.[10]

In the arena of medieval discourse, it is not clear whether derogatory language about Jews would have struck people as uniquely hateful. Anna Sapir Abulafia has noted that the language directed against Jews in the twelfth century was part of a general discourse of abuse: "Accusations of carnal or unreasonable behavior were limited neither to Jews nor to the twelfth century. Christians would happily employ similar terms of abuse against those with whom they disagreed on all kinds of issues. Invective was very much a commonplace in the disputations of the eleventh and twelfth centuries."[11] Can we trust the most extreme and violent language that Christians used about Jews to represent a permanent system of beliefs and attitudes? Even more crucial, perhaps, is the question: In an environment of inflated rhetoric, how seriously did people take this language attacking the Jews? Did anti-Jewish polemic automatically gain acceptance among Christians and galvanize anti-Jewish violence? Jews, as well as Christians, must have learned to assess the language applied to Jews in the context of this violent style of discourse.

The language of the church had always oscillated between references to the protection of Jews and their condemnation. Even popes who indulged in violent language against the *perfidia* of the Jews reissued bulls protecting Jewish property, guaranteeing their rights to worship, and insisting on their freedom from forced conversion.[12] It is difficult to draw conclusions from the repeated reissuing of the protective bulls and their accompanying anti-Jewish language. The papal commitment to the protection of the Jews may indicate the ability of certain

Jewish communities to purchase papal patronage, the effectiveness of which was idiosyncratic. Clearly some clerics were radicalized and devoted enormous energy to anti-Jewish efforts.[13] None of these efforts, however, fundamentally overturned the rhetoric of tolerance. Even if that papal rhetoric coexisted with anti-Jewish feeling, the survival of the *sicut judeis* tradition through periods of persecution and inquisition is quite remarkable. (It remained a central theme of papal policy until the years of the Counter-Reformation and the imposition and advocacy of harsher measures against Jews.)

Another question to consider is how responsive were Christians to this defamatory language about Jews. Most of the texts and narratives that recount the various blood libel accusations probably circulated only among a literate clerical elite. They surely tried to transmit some of their fantasies through sermons or drama or art to the laity. Because these clerical statements are usually our only source for the internal thinking of lay people, we may have been too quick to assume that a lay audience automatically believed a preacher and internalized his message. Certainly, many lay audiences could be suspicious of preachers and their messages, as the various medieval anticlerical heresies and sentiments suggest.[14]

Christians did not always receive clear messages about Jews. Did Christians, for example, believe the occasional radicalized friar or the long succession of papal pronouncements protecting Jews? Even depictions of the Passion provided competing images that shifted polemical focus away from Jews. The internal development of artistic representations of the Passion, for example, suggests that the Jewish role was often blurred or obscured. A recent study of anti-Jewish images in Christian art has argued that increasingly historicized Passion scenes in the late Middle Ages made it harder to single out Jews in the crowded images of soldiers and bystanders. The putatively historical clothing would also have made it difficult to connect the figures to contemporary European ethnic groups. At the same time, the symbolic nature of the Passion as speaking to the general condition of man was powerfully represented without a specifically anti-Jewish element. In this view, "the abhorrent, satanic ugliness of the catchpoles and the crowd, often caricatured to the grotesque and in sharp contrast to the features of Jesus, is a timeless image of Man at the extremes of his nature rather than an accusation against a specific group such as the Jews or the mercenaries."[15]

Some Christians were clearly affected by the language and violent images, but many others were not, or at least did not see these images of Jews as the only way to define these people. Jews themselves must have been aware of the ability of Christians to discriminate among these images and added this knowledge to their own decision making about how best to protect themselves and their families. Taking all these considerations into account restores to medieval people on all social levels a degree of choice and autonomy in how they responded to Jews—and how Jews responded to expressions of Christian anti-Judaism.

※

Rhetorical violence against Jews looks very different when seen in the context of larger polemical trends in medieval culture. The long-term significance of physical violence against Jews might appear different as well if we look at the evolution of violence rather than focus on particular episodes. The progress of real animus against Jews was incremental and idiosyncratic. The nature of violence against Jews suggests more about the resilience of local Jewish communities and the contingency of anti-Jewish actions than it does about an inexorable movement of anti-Judaism. Violence against Jews, we will see, has more the character of disconnected outbursts. It is hard to see that these episodes had a long-term impact on the way Jews and Christians treated each other in routine situations.

Despite the church's traditional position protecting Jews from forced conversion, some violence, as we have seen, was generated by aggressive churchmen or violent nobles in the early part of the eleventh century. The transience of this violence suggests that Jews were already firmly integrated into these different locales. Having survived the period of feudal transformation relatively unscathed, the Jews also weathered episodes of violence later in the eleventh century, including the infamous attacks in the Rhineland, which did not fundamentally alter the nature of Jewish life. It is for this reason that, Robert Chazan concludes, "1096 did not usher in a period of unrelenting insecurity for Jewish life. When the violence of the spring months subsided, much of Jewish life returned to the status quo ante. As we have seen, the Second Crusade brought with it little actual violence against European Jewry. The slanders of the twelfth century occasioned anxiety and fear, but little bloodshed."[16] The Jewish communities at the center of the violence were quickly rebuilt, drawing on increasing immigration and an expanding twelfth-century economy in northern Europe.[17]

Even when the social disorder of the Crusade attacks passed, Jews had to confront accusations of child-murder and ritual crucifixion. Some of these accusations led to violence against Jews. A brief survey of these accusations reveals the idiosyncratic, random, and temporary nature of the effects of such incidents on local Jews. So many of these episodes depended on the quirks of religiosity of individual rulers or clerics that connecting them creates an arbitrary and deceptive impression of a linked evolution of anti-Jewish sentiment. Recognizing this makes it more difficult to see accusations of ritual murder as expressions of endemic anti-Judaism that would ultimately lead to bloodshed. For many people involved in the creation and promotion of the accusations and the associated martyr-dom, the creation of a local cult may have been less about demonizing Jews than about stimulating a specific kind of Christian communal piety.[18] It is hard to imagine that outside of the moments of accusation, most of which were defused by royal authority or local indifference, the creation of ritual mur-der accusations had a lasting impact on how Jews perceived their long-term security.

The most important factor in the limited popular response to these first accusations was the reaction of royal and local agents. In England, for example, royal authorities intervened quickly. In Lincoln in 1202, Stafford in 1222, Winchester in 1225, and London in 1244 and in 1250 accusations against Jews were short-circuited by the royal judicial system (also without provoking popular outrage or "mob" violence). Imprisoned Jews were freed, no shrines were established to the alleged vic-tims, and there were no further local accusations of ritual mur-der.[19] Again, the aftershocks from these outbursts quickly dissi-pated. It took the intervention of Henry III, a self-consciously pious ruler, prompted by a local knight intent on incriminating Jews, to bring real violence to a Jewish community following an accusation. In 1255 the king inserted himself in a local case and triggered the judicial murder of some thirty Jews. Before Henry's arrival, local authorities had taken no action against Jews, nor had there been any popular attacks.[20]

Accusations were idiosyncratic and seemed disconnected from mainstream religiosity. The 1144 accusation in Norwich was the creation it seems of Thomas of Monmouth, a middle-level cleric looking for a way to bring a saint's presence to his local church. His story of the murder of a young Christian boy by local Jews found little sympathy or audience in Norwich.[21]

According to a recent study of the rhetoric against Jews, the erratic level of offerings to the shrines associated with ritual murder cases suggests that the shrines made very little impact on popular piety. By 1341, for example, annual offerings at William of Norwich's shrine in England had dropped to 6d.[22] Such indifference to the shrine must reflect some disregard as well for the story of the saint's purported demise. It suggests that local Christians—both clerical or lay—did not believe the accusations or that even if they did, they were not so central to local religious identity that they had to make those beliefs manifest either by prayer at the shrine or by seeking vengeance on Jews.

The next accusation of a similar crime appeared in 1168 when Jews of Gloucester were charged with crucifying a young boy.[23] It is not clear whether there were any repercussions from the accusation. A shrine was established at Paris sometime between 1163 and 1171, but what happened to the local Jewish community is unclear.[24] It was not until 1171 that Stephen, count of Blois, acted on a similar story, burning thirty-one Jews for allegedly killing a local child. It is worth noting, too, the very limited impact of Stephen's attacks beyond this first group of victims. No shrine was "created" to the boy to memorialize the incident, and once the display of comital power was over, the episode had no religious resonance for the local population.[25] Indeed, it seemed very difficult to create shrines to these putative saints. In 1181 one such shrine was established at Bury St. Edmund's, but in 1183 in Bristol and Winchester in 1192 accusations against Jews dissipated with no lasting memorial. In fact, until 1244 no new shrine was established although periodic accusations surfaced.[26]

The long-term numbers of host desecration accusations are also suggestive that many Jewish communities must have felt themselves relatively secure from such outbursts. One tally of the numbers of accusations records fifty accusations between 1230 and 1566, with thirty coming between 1290 and 1400, and of those only fifteen resulted in some kind of violence against Jews. In the fifteenth century there were ten episodes, of which six ended in rioting. Jews would have been aware of such accusations, but many must have lived their entire lives without direct experience of those accusations.[27]

Even when violence against Jews was not associated with ritual murder accusations, attacks against Jews often were unexpected outbursts related to local events. Christian chroniclers themselves were shocked at the outbreak of such violence at King Richard's coronation. Richard quickly acted to stop further attacks.[28] Other attacks against Jews at the end of the twelfth century, including the large-scale massacre at York, were the product of crusading fervor. Jewish wealth—acquired through the dubious practice of money-lending—was increasingly seen as a legitimate target for kings or pious crusaders as a way to fund crusading. The attacks against Jews were also fired by frustration with the policies of Henry II and Richard who had started collecting the debts owed to Jews for the royal treasury. Richard's political opponents also used the attacks as a way to strike at his authority.[29] In York for example, it was clear, at least to contemporaries, that local barons instigated the violence against Jews.

In Richard of Newburgh's chronicle account, it was not hatred of Jews as a religious group, but political conspiracy galvanized by "certain persons of higher rank who owed large sums to those impious usurers, the Jews."[30] Newburgh condemned the perpetrators of the crime: "Their first crime was that of shedding human blood like water, without lawful authority; their second, that of acting barbarously, through the blackness of malice rather than the zeal for justice; their third, that of refusing the grace of Christ to those Jews who sought it."[31]

However much Newburgh may have disliked the Jews for their usury, his comments made it clear that they were not facing anti-Jewish sentiment throughout Christian society. In these cases Jews were victims because of their actions and their association with the crown, not simply because they were the targets of a constant religious animus.

⚙

The record of violence against the Jews in French lands of the thirteenth century is equally difficult to use as proof of the marginalization of Jews. In the first place, Jews in France lived in very different kinds of societies. Life in the south of France often has been characterized as having the quality of a more Mediterranean *convivencia*. The image of a tolerant multicultural south is one that began in the Middle Ages by southerners looking nostalgically back at a world they thought destroyed by the northern French crusade against the Cathar heretics of the south in the thirteenth century. Even when taking such nostalgia into account, for the Jews—at least the vast majority living in the large cities of the coast—the society of the south did allow a more variegated Jewish community to develop. Jews owned land; they marketed grain, livestock, wine, and meat. They labored as doctors, tailors, skinners, leather merchants, masons, manuscript illuminators, and dealers in old clothes.[32] In all of these roles, they engaged in business with Christians as well as other Jews.

No doubt the increasing northern anxiety about heretics and royal impatience with the laissez faire culture of the south unsettled Jewish communities. However, the upheavals and violence of the northern crusade against the Cathars seems to have left little impact on Jewish communities of the south. It is interesting that no major massacres of Jews (except at the bloodbath in Beziers when the entire town's inhabitants were slaughtered by crusading armies) occurred during the long years of conflict. No doubt Jews suffered as part of the general population when the sieges and battles affected their homes, but there is no evidence that the great Jewish communities of Toulouse, Carcassonne, and Narbonne suffered specifically anti-Jewish attacks.[33] Even with the caveat that most of the violent conflict against the heretics occurred away from the centers of Jewish population, Jews did not become a special target of the northern French armies, nor were they treated as scapegoats by the defeated southerners.[34]

The first example of what seemed to be mass violence against Jews occurred when crowds of so-called Shepherds and laborers rioted to protest the capture of Louis IX by Muslims in Egypt during his failed attack on Damietta. In 1251 these mobs found some support in various French towns but even Blanche, Louis's mother and regent, turned against them. From the few records we have, it seems they vented their rage not just against Jews but clergy and the wealthy as well. To account for their anti-Judaism as mirroring the well known anti-Judaism of the king does not explain their simultaneous animosity to the clergy or the rich, sentiments that Louis would surely have opposed.[35]

The reappearance of violent groups of Shepherds coincided with the announcement of a crusade following the famine of 1315–17. In 1320 a large but unknown number of people—the majority from marginal social groups—migrated to Paris in search of the king's leadership on a crusade. Philip V ignored them. After attacking the royal prison, they moved south into Aquitaine. Along the way, they attacked Christians who drew their ire for their wealth or social authority. Attacks on Jews were recorded in Saintes, Verdun, Cahors, Toulouse, Albi, and several other towns as these mobs moved to various regions in the south.[36] The violence in Toulouse seemed particularly brutal.[37] In response to the growing disorder, the *sénéschal* of Carcassonne dispersed a large group of these so-called Shepherds across the border into Gascony, Navarre, and Aragon. The attacks by the Shepherds are problematic measures of "popular" anger. This was not institutionalized persecution that demanded widespread collusion but was the product of violent individuals who had shaken off the normal constraints of medieval society. The royal government may have failed to protect Jews from the Shepherds' attacks, but they quickly moved to restore order and exact retribution for the violence.

Once the outbursts by the Shepherds had passed, it is likely that Jews did not imagine that such violence would recur. The Jews in the southern territories ruled by the Capetians had been under steady financial and regulatory pressure from the French monarchy, but these local communities of Jews had no reason to expect continuous eruptions of uncontrolled violence.

Unfortunately, a rumor that Muslims had conspired with Jews to have lepers poison wells circulated in 1321. Lepers were attacked and Jews suffered in many episodes of mob violence. Our task must be not only to record the particular events of this outbreak but to contextualize them as well. Historians have disagreed for generations on the numbers of victims. In Chinon, for example, the estimates range from 8 to 160.[38] Jewish communities were still able to provide a huge sum as a fine for their alleged involvement in the Leper plot, suggesting that the violence of the previous years had not destroyed their economic resources.[39] The record of violence against Jews in French lands is thus difficult to interpret. Did Jews in fourteenth-century France, after their readmission in 1315, feel they were living in a more dangerous and unstable world? There were clearly outbreaks of Christian cruelty and fanaticism. At the same time, Jews did not willingly relinquish their places in local societies once they saw that the violence had passed.

⊠

Jews in thirteenth- and fourteenth-century German lands certainly endured more violence than their brethren in England and France.[40] The early years of the twelfth century gave no indication, however, of increased violence. After the attacks in the Rhineland during the First Crusade, the records are silent for almost fifty years about threats to Jews. The Second Crusade brought warnings of violence and small-scale attacks against local Jews, sparked largely by the preaching of a Cistercian monk with a particular animus toward Jews. Local Jews were protected by bishops and the king when mobs and crusaders turned violent. Significantly, the great spokesman for the Crusade, Bernard of Clairvaux, made it clear he would not tolerate any attacks on Jews. Still, some Jews suffered, but historians agree that the violence never reached the proportions of 1096.

In the remaining years of the twelfth century, Frederick Barbarossa's rule was strong enough to control most local disturbances against Jews. That stability survived into the thirteenth century as well. The very vitality of Jewish communities must have provoked the complaint by Pope Gregory IX to German bishops that Jews were too powerful in Christian society. The position of Jews in German towns was only occasionally threatened in these years. For example, a ritual murder accusation appeared in Erfurt in 1221.

In 1235 in Fulda, a particularly vicious change occurred in the nature of accusations against Jews. Now a charge of ritual cannibalism was added to the accusations of child murder or crucifixion.[41] Whatever the exact process or evolution of the slander, Frederick II suppressed the accusation at Fulda, but it clearly survived in some clerical circles. The accusation entered the repertoire of anti-Jewish images at the disposal of Christians. (However, it is interesting that it was never raised during the plague or later accusations of host desecration.) Despite the occasions of violence, Jews still felt free, as Rabbi Meir b. Baruch of Rothenburg said: "The Jews are not subjugated to their overlords as gentiles are ... their status in this land is that of a free landowner who lost his land but not his personal freedom."[42]

Serious but isolated attacks against Jews occurred in the second half of the thirteenth century in Frankfurt in 1241, and in several other towns a year later. Some of these may have been associated with trials arising from ritual murder accusations. A judicial attack clearly occurred in 1244 when many Jews in Valréas were executed after being convicted of killing a Christian girl. Even this kind of trauma did not leave other Jewish communities helpless or shaken. They appealed to Pope Innocent IV, and in lieu of imperial protection, he issued papal bulls to crush the ritual murder accusations.

In the following years, Jews were perhaps more vulnerable because of the absence of a strong imperial ruler. Nevertheless, most towns and regions continued to treat Jews with respect and offered protection, even if serious attacks on local Jews occurred in several towns. By the end of the century, mobs attacked Jews in Mainz in 1281 and in 1285 in Munich, where it seems that close to two hundred Jews were burned in the synagogue. Both attacks were linked to the murders of children. City councilors and local church officials tried to control the violence or exact vengeance on the Christians who could be apprehended. The level of violence must have affected the sense of security of some Jews, for a small group including prominent rabbis tried to leave Germany. The king responded by restricting Jewish travel. By 1287 the ability of the king to protect

Jews was increasingly ineffective as a wave of attacks in the Rhineland spread through twenty towns. Hundreds of Jews were killed.

The worst period of violence against Jews began at the end of the century. The infamous Rindfleisch massacres began in 1298. The first explosion of violence occurred in Rottingen in April 1298 and continued in waves over the following two years. The violence, usually associated with lower-class mobs, spread throughout Franconia and into Bavaria. Recent work estimates the destruction of 146 Jewish communities and the deaths of at least three thousand people.[43] In the context of medieval evidence, establishing the numbers killed is often difficult, but it is clear these were major attacks. As with the period surrounding the Rhineland massacres of the First Crusade, the normal structures of political authority had been disturbed. A struggle for the imperial throne in the late 1290s apparently allowed for the buildup and then explosion of social tensions directed toward Jews.[44]

Although the severity of the violence tempts one to explain its origins in fevered Jew hatred, it is difficult to be certain of the motives of all the attackers or the sequence of events that triggered the violence. Some observers felt that greed was involved as much as piety. Others expressed their confusion and uncertainty over the motives of the perpetrators. The situation was more complicated and nuanced than simply mass vigilante or mob attacks against local Jewish populations. It is interesting to note that in a couple of towns church authorities were skeptical of the original charges of host desecration that seemed to provoke some of the attacks.[45] (In the fourteenth century in Passau, chroniclers would blame indigent priests for provoking the host accusations.)[46]

Even in the midst of the violence, Jews were not completely abandoned. We have several examples of attempts to protect Jews—even if these were condemned by the chroniclers. For example, Rothenburg townsmen helped Jews escape.[47] The residents of the town of Wildstein, where refugees from the attacks in 1298 gathered, tried to help Jews before the town itself fell to the mob.[48] The situation in Nuremburg was even more dramatic. The mayor and the castellan gave Jews refuge in the castle, but townsmen turned on the Jews and killed them. The king and royal officials did not let this go unpunished. The king fined the town and banished the perpetrators.[49] When the attackers reached Regensburg and Augsburg, the towns were able to fend them off, thus saving their Jews.[50] Royal power was not complicit in the attacks, and although it offered little resistance to the attacks as they progressed, it acted swiftly to punish the guilty. Imperial policy toward Jews in the early years of the fourteenth century evolved into a mixture of measured extortion and official protection.

Attacks on Jews had become an accepted way for marginal elements in German society to vent their frustration. The period 1336–38 was marked by the most widespread attacks on Jews beginning with the attacks by a poor knight near Würzburg, who was in debt to Jews. Seizing on a rumor of Jewish desecration of the sacrament, this Armleder, so called for his leather arm guards, launched a series of attacks on Jews in the southwest of Franconia. The Armleder attacks, were, according to Miri Rubin, "more brutal" than previous violence and involved larger groups of people, including knights and their retinues. The target of the attacks, however, may have been broader than just Jews. One chronicler asserted that the knights had attacked not only Jews but others who seemed to be parasites on society, including bishops, clerks, monks, nuns, and scholars.[51] The authorities slowly responded to the threat to social order. The first Armleder leader had been captured and executed in 1336. It took the combined efforts of lords, clergy, and towns to crush the last appearance of the Armleder groups in Alsace by 1338. Others took up the banner in Bohemia and Austria, but the violence quickly faded. The efforts of imperial or ducal authority to suppress the attacks, even if they largely failed, suggest as well that the mania against Jews had not overtaken everyone.[52]

It is hard to tell if Jews returned to all of the towns that had been targeted by the Rindfleisch and later by the Armleder bands. Jews certainly stayed in towns that protected them such as Augsburg, suggesting that they still relied on their experience of immediate local conditions to make decisions about their best chances of safety.[53] Again, we are left with the mystery of how Jews can become the despised victims in certain locations and remain protected members of local communities in others. Even these concentrated attacks could not eliminate Jewish communities. The number of Jewish settlements had risen by the time of the plague in 1348 to over a thousand from about fifty in the middle of the twelfth century. Even after the severe violence of the early years of the fourteenth century, the Jews chose to live in small-scale Christian towns or villages, which dominated the German landscape.[54] Such an increase in the number of settlements argues against the conclusion that Jews simply could not leave German lands. The violence of the late thirteenth and early fourteenth centuries did not disturb the general pattern of Jewish settlement.

The resilience of Jews and their attachment to places they considered to be their homes can be seen again during the accusations and attacks of the plague years around 1348. The plague brought a demographic disaster to most regions of Europe. The trauma of mass deaths destroyed the social fabric of many towns and villages. In some regions in Germany Jews were at-

tacked as potential carriers of the plague, but Christians imagined many other explanations for the plague as well.[55] In the
general social unrest of the period, it may be difficult to understand the precise evolution of the violence. The attacks often
may have been the result of local social and political conditions
unrelated to the specific outbreak of the plague.[56] Certainly, the
weakness of the king, a factor in almost all the cases of assaults
on Jews, was evident in the years of attacks on Jews associated
with the plague. Whatever the cause or trigger of the violence,
it touched a section of the Jewish population. Records tell of
Jews killed in most German towns, although where the king
ruled directly, in Bohemia and the Habsburg regions of Austria,
Jews were protected. Regensburg also saved its Jews.

However, what needs to be emphasized is the aftermath of
the plague-related violence. In virtually every case where a
community was struck by violent outbursts, even to the point
of dissolution, within a few years Jews had reentered the towns
and refounded communities. Towns in the Palatinate, for example, actively attracted Jewish settlers even in the immediate
shadow of the plague.[57] Jews quickly returned to communities
in Swiss territory where they had been driven out by plague
massacres.[58] Jewish communities in the Low Countries also survived these years without major traumas from attacks.[59] Very
few Jews were killed in Poland or accused of involvement in the
spread of the plague.[60] Some communities in Germany escaped
untouched by plague-related violence. Ratisbon town councilors, for example, protected their Jews from the Rindfleisch
and Armleder attacks as well as later outbursts during the time
of the plague. According to one contemporary account, "The
burghers of Ratisbon, wishing to honor their city, prohibited
the slaying or destruction of Jews without judicial judgment.
They said that if God himself should desire the death of the
Jews, they would not resist; but they wished first to achieve

greater conviction that it was God who imposed that penalty upon the Jews. Thus the Jews of Ratisbon, though not without much difficulty, have hitherto escaped death by burning."[61]

Even where we know Jews were murdered—as in the case of Königsburg—they were readmitted almost immediately after the attacks.[62] Erfurt and other towns saw immediate Jewish immigration after plague attacks.[63] Nuremburg, too, readmitted the Jews, by order of the emperor after hundreds were killed in plague massacres.[64] In Freiburg the city council apparently tortured Jews and extracted confessions that they had poisoned wells in the region. Jews were burned—except for pregnant women and children. Still, Jews returned to the town under the protection of the Habsburgs in the second half of the century.[65] It was difficult to dislodge the Jews from the localities they knew. As Alfred Haverkamp concludes, "One of the many new findings of the *Germania Judaica* is the observation that in the majority of the approximately 500 locations where Jews resettled after 1349, often after an interval of more than a decade, they generally relocated to places they had previously occupied."[66]

Jews who survived the various outbreaks of violence in the first half of fourteenth-century Germany surely looked on the gentile world with suspicion and likely moments of true hatred. Nevertheless, they did not permanently abandon those towns and villages that had been their homes. The violence itself or the memories of that violence were not enough to completely undermine the connection of Jews to these places. We do not know, of course, what Jews felt when returning to a town or village where they had seen friends, family, or neighbors killed. Nor for that matter do we know what Christians thought on their return. If actions speak louder than words, then we have to take seriously the intentions of Jews who remained engaged with gentile society in fourteenth-century Germany. Their resettlement in no way indicated forgiveness or historical amnesia. However, they must have felt that they could rebuild their lives and look to the future.

⊗

It is in the thirteenth century that the history of Jews in Spain became marked by the persecution and uncertainty more traditionally associated with France, England, and Germany. Pressure for conversion was brought to bear from mendicant preaching, staged disputations such as the 1263 Barcelona debate, and the production of explicitly anti-Judaic literature. The course of the fourteenth-century Shepherd attacks in northern Spain was similar to the pattern of local explosions of violence followed by a return to stability and the status quo that we saw in France. It is striking how limited the violence was. The monarchy clearly had no desire to exploit the anti-Jewish agitation. Indeed, it resented the intrusion of these anarchic groups. After an attack against the Jews of Montclus (although the extent of that attack remains uncertain), royal authorities crushed the Shepherds, and by July there was no further violence against either Muslims or Jews. Local records contain no other evidence of attacks beyond this one town.[67] The efforts of the monarchy to recover Jewish property—and thus royal income—apparently provoked some locals to resent Jews. Nevertheless, Jews who lived in the region were not intimidated. After coming to the town to bury the dead, Jews vented their own rage in a riot. These were not people cowed by the recent attacks. They wanted revenge.[68]

Jews faced other situations of violence in Spain in the early years of the fourteenth century. In some of these cases, as in Girona and Valencia, the violence—a kind of ritualized social disorder—had clear limits and did not fundamentally disturb Jewish life.[69] This kind of disruption and later accusations of well-poisoning in Torrega surely gave a painful frisson to daily life, but the point is that Jews—as with many other groups—were used to such tension; they clearly had learned to live with it and endure its uncertainties.[70] The traumas of the plague years of 1348 and its aftermath, and then the long civil war in Castile when Henry of Trastamar challenged his half-brother Peter for the throne, left their marks on the local Jewish communities, but the fundamental equilibrium of Jewish life was not permanently disturbed.[71]

The infamous 1391 riots have been understood in much the same way as the First Crusade attacks in the Rhineland—to signal the beginning of the end of Iberian Jewry. The riots created the huge *converso* population that later sparked the royal policy of expulsion and inquisition.

Evidence that the motivation and impacts of the attacks on Jews was complex undermines the view that these blows were only an expression of anti-Jewish sentiment "simmering" beneath the surface.[72] In an older but still relevant article, Philippe Wolff has encouraged us to see the 1391 violence in the context of other late medieval class rebellions that exploded during the preceding decade in Florence in 1378 and in 1382 in England and elsewhere.[73] The Jews clearly occupied a generally favorable position in the economic hierarchy of the towns. They were an easy target for lower-class urban people of the Spanish towns when those individuals became radicalized, desperate, and violent.

The particular social conditions preceding 1391 created a loosening of royal authority. Juan I of Castile, who had urged moderation on the vocal anti-Jewish archdeacon of Ecija, Ferran Martinez, died in 1390, leaving a minor as heir.[74] From that point on, local authorities seemed largely unable to control the crowds of laborers and other lower-class urban groups. Violent attacks on Jews spread from Seville to Barcelona, and it was only after two months or more in the late summer and fall that concerted actions by royal officers crushed the attacks. Despite the extent of the violence, Jews were not completely abandoned. Royal officials often put themselves in harm's way to protect Jews. When official forces mobilized, they sometimes were able to short-circuit the violence as at Saragossa and Heusca.[75]

For Jews in the aftermath of the attacks there seems to have been tension between the desire to preserve the continuity of their lives and difficulty in overcoming the trauma of the attacks. At Palma for example, half of the converts wanted to resettle in the local Jewish neighborhood. Jews in other places, most notably Barcelona, could never reestablish themselves into viable Jewish communities. Despite the uncertainty over how to restart communal life, Jews had every reason to think that the authorities would not tolerate any similar kind of outbreak. Authorities in both Aragon and Catalonia had made clear their hostility to the murders and to the forced conversions.

Among Jews who survived the attacks, and probably among Jews whose communities were untouched by violence, there must have been a sense that—at least for a time—the fundamentals of their world had shifted. In terms of the numbers of the victims and the demographics of the surviving communities, it was a time of retrenchment. By 1415 at least one-third to one-half of Iberian Jewry is thought to have converted to Christianity. In the large Castilian cities, the Jewish communities never recovered. Many Jews moved to smaller villages and towns, perhaps hoping they could assure their safety better than in cities where mobs could form quickly. Many Jewries did not recover; only eight were left in Catalonia, five in Valencia, and twenty-two in Aragon.[76] On the other hand, Jews in Portugal and Navarre escaped any outbreak of violence. This very quick rehearsal of events in 1391 does not, of course, do justice to the widespread trauma of the violence and forced conversions. It would be doing a disservice, however, to the complexities of the medieval Jewish experience to make 1391 a sign of the inevitable degradation of the Jewish community that would then be finally effaced in 1492. The very possibility and relative success of such widespread conversions suggests, of course, that a real *convivencia* had existed before the riots.

Benjamin Gampel's reflections on the problem of post-1391 life are insightful:

> To some students of Iberian Jewry, especially those who study the last century of Jewish life on the Peninsula, the notion of *convivencia* seems particularly problematic. For Iberian Jews, after all, this was a century marked by ruthless pogroms, massive forced and voluntary conversions to Christianity, and increasingly restrictive legislation culminating in expulsion. Where does one find tolerant "living together" in this bleak century? And yet, we cannot overlook the relative social and economic health of the Jewish communities over the course of the fifteenth century, nor can we pass over—as too many scholars have done until recently—the striking and original intellectual contributions made by Jews during these years. Even in the very last years of the Jewish life in the Peninsula, the picture is not as gloomy as one might have thought. When Ferdinand and Isabella acceded to their thrones, it was still not obvious that expulsion was inevitable; certainly the term *convivencia* could still be applied to this period with some justice.[77]

Gampel is surely right. Even a cursory glance at the conditions of Jews in Spain after 1391 suggests that violence and conversions did not destroy Jewish life or even mortally cripple it. Smaller-scale communities of Jews continued to thrive in Spain, and if we can trust the numbers at the time of the expulsion, some kind of demographic growth must have occurred to rebuild the community. The Jews in Morvedre, as we know from Mark Meyerson's brilliant reconstruction of their lives after 1391, remained highly integrated into the local economy and society. They even made efforts to preserve connections with the new *conversos*.[78] Indeed, the 1391 attacks had not created a situation where anti-Judaism was rampant. In 1454, for example, there was an acquittal in a ritual murder case in Valladolid. At the same time, personal relations between Christians and Jews could thrive as they did when a Jew worked with two Christian friars on a translation of the Hebrew Bible for a Christian patron.[79] The Jews lived for a century in Spain after the 1391 attacks. Many Jews must have continued to feel connected and integrated into their local worlds. For the generations that grew up after 1391, it was still the only world they knew.[80] By the second half of the fourteenth century, then, many Jews may have lived in increasingly perilous and tenuous relations with Christians. They lived with memories and reports of traumatic violence and social disorder. But as we have seen in the previous chapters, the Jews seemed ultimately to be local people who continued to build thriving communities in Spain.

Does studying violence against Jews in this period help us understand the fundamental experience of Jews in medieval Europe? We certainly can see how vulnerable Jews were at many points and in various locales. But important questions remain if we are to get at the essential experience of being a Jew in medieval Europe. Were they more vulnerable than other groups of people who became targets of different kinds of violence? Did they see violence as a permanent threat that colored their everyday lives? When violence erupted did they envision the end of their communities or did they see it as a gauntlet through which Jews had always managed to pass? Did anxiety about violence from Christians routinely affect relations with Christians or only during periods of open disorder? Did they see every Christian as a potential enemy? And how did individual Christians respond to Jewish neighbors and business partners? Since we cannot know the experience or thoughts of each individual Jew or Christian in medieval Europe, we are left with trying to gain impressions of how groups or communities responded. Our generalizations are only based on moments in time, often during periods of great stress. Still, the image of a constantly beleaguered and victimized Jewish community is surely something that must now be measured against the resilience of Jewish life. If our understanding of violence against Jews needs to be revised, so too does how we fit the expulsions of Jews into our narrative.

6

Expulsion and Continuity

Although Jews survived periodic outbursts of violence in the Middle Ages, they were not able to escape the expulsions ordered by most European monarchs by the end of the fifteenth century. The expulsions have provided a convenient terminus to shape the narrative of the Jewish experience. They have served as the natural end to a story of persecution and suffering. England expelled its Jews in 1290, and France's final expulsion of the Jews occurred in the late fourteenth century. Jews were forced out of Spain in 1492 and then a few years later from Portugal. The expulsions apparently confirmed that Jewish residence in western Europe was conditional and temporary. Recent scholarship on England and France has made a persuasive case that while multiple factors contributed to the expulsions, key elements were the mendicant-inspired piety and anti-Judaism of monarchs. Both English and French rulers preferred to acquire what seemed reliable grants of taxation from their subjects, in return for expelling Jews, rather than relying on money expropriated from Jews. That process was often arduous since the funds could be seized only if and when the king enforced the payment of debts to Jews. Subjects who agreed to the grants in return for expelling Jews may well have imagined a future free of debt to Jews without considering that the monarchy would still try to collect those debts. Albeit imperfectly, the expulsions seemed to address the economic needs of kings and subjects.

I cannot add very much to the work of previous scholars on the expulsions themselves. What I can do, I hope, is suggest that the expulsions are not teleologically linked to the preexpulsion experience of Jews in England and France. The trigger for expulsion seems to have been decisions taken by individual rulers at moments of crisis. King Edward of England, for example, recovered from a grave illness and was frustrated at the failure of his crusade enterprise. He turned to the Jews to vent his anger. A similar pattern can be seen with Louis IX who was frustrated with the Jews' refusal to convert and his own defeated crusade endeavors. The crucial role the monarchs played in making the expulsions happen needs to be emphasized because it is easy to imagine the expulsion as the expression of an entire Christian society. While there may have been some support for the expulsions, particularly from those groups feeling the burden of the debts to Jews most acutely, the expulsions were still the actions of monarchs, carried out by their ministers and servants. The expulsions did not come in response to a popular call for the Jews' removal. In fact, when nonjudicial violence flared against Jews in late-thirteenth-century England and France, the authorities usually quashed it. Monarchs did not take the opportunity of the violence against Jews to expel them. The decisions came at moments of relative social peace and stability.

As it became clear that Jews, despite the restrictions placed on them, would not convert, expulsion seemed a practical alternative, particularly when it would secure a money grant to the monarchy from beleaguered debtors. What the general population thought about the expulsion is hard to know. Whatever the exact feeling of Christians at the time, we can assume that Jews faced trauma and distress at being uprooted from their homes, the only homes many of their families had known for centuries. Indeed, it is clear that some Jews were not willing to accept their loss of "country." They did not see their expulsion as necessarily permanent and irrevocable. We know, for example, there were Jews still living in Gascony, from which the English king had expelled them in 1287, as late as 1308.[1] Even though this seemed to anger English authorities, it is not clear that they could completely enforce the expulsion. In France, of course, Jews continued to return whenever they were offered readmission. Moreover, some Jews did not want to abandon their connection to England even after 1290. Jews explored returning to England in the early fourteenth century and made the first attempts at trying to negotiate a reentry in 1310, but the monarchy did not wish to reopen the question of Jewish settlement.[2] The expulsion from England should not be seen as the inevitable purging of non-Christian elements from English society, but as an experiment in crisis management over taxation. There was no sorrow that the Jews had left, but we have to avoid reading a unified "national" consensus into the event. Some Jews did not view the expulsion as the final break that it would become with England. They could imagine returning to England, which suggests, of course, that they likely understood they would face hostility and obstacles. Nevertheless, they must have felt that English society was not intolerably hostile to Jews. They had been expelled from various English towns and had always found another refuge and settlement. It is logical to assume that they imagined the passions of Edward would eventually pass as they had for periodic expulsions of French Jews earlier in the century.

The history of expulsions from France is particularly complex, making the expulsions difficult to see as part of a cohesive policy toward Jews. In fact, the repeated expulsions and then readmissions should suggest that French kings were uncertain about how to treat Jews and that the decisions to expel and readmit Jews were based on pressing financial and political issues as well as royal religious prejudices. They did not represent the fulfillment of a long-term French policy.[3] It is difficult to deny, of course, that royal aggression had degraded the position of Jews in French society. In the years after the readmission in 1315, the Jews had become increasingly marginalized, although the crown made efforts to protect them from rapacious officials and local extortion.[4] The Jews had yet to lose faith in France. The repeated expulsions—even if not the result of a consistent, long-term policy—the takings, and associated violence had winnowed down the size of the community. (However, at least with the major expulsion of 1306, the wealth collected by the crown suggests that Jews had remained a vital and economically powerful community).[5] Jews had in the end become foreign merchants, who were soon to be finally expelled in 1395.[6]

The French expulsions have been understood as an expression of an increasingly centralized monarchy that during the course of the Middle Ages saw France as a chosen nation destined to lead Christendom. The Capetian kings of France and those who celebrated the kings presented France as increasingly obsessed with the religious and social purity of the kingdom—a kingdom that by its very nature could not tolerate the presence of Jews.[7] Using this as the basic framework for a "master narrative" of the history of the Jews in France, the goal of most recent scholarship on medieval France and the Jews has been to explain the end of the Jewish community in France. The picture we have is of an inexorable squeezing of the Jews out of the borders of Capetian France.

The expulsion in 1322 has been described recently as the king's "way of assuaging popular discontent at the governance of the state, its failure to bring the blessings of heaven to earth."[8] Perhaps. But would the polity have really been appeased by the expulsion of the small number of Jews remaining? Was the crown thinking in this way—that it had to be so responsive to putative popular concerns? Moreover, this interpretation assumes a view of the French state that may be overstated. The king and some of his advisers may have internalized the idea of the Holy Land of France and the French as the new Chosen People. It is far from clear that the majority of French subjects would have thought about their king and state in this way—and thus made the connection with expelling Jews as a way to purify the kingdom. It may be that medievalists have been too anxious to anticipate the formation of a true nation state—a mentality that seems to have been activated only later in the ongoing struggle of the Hundred Years War.[9] Whatever the final motivations of the crown, the few Jews remaining in 1395 were expelled.

The problem with linking the particular expulsions of England and France to Spain, of course, is the huge range of time that intervenes. The expulsion from Spain did not take place until two hundred years after the Jews' removal from England. The differences in time, culture, and geography make it difficult to treat the expulsions as common phenomena or evidence of similar societal attitudes toward Jews. It may be impossible to explain why the expulsion from Spain took so long to occur. Spain was equally exposed to the rise of mendicant piety, and increasing pressures on the Jews in the thirteenth and fourteenth centuries did not lead to large-scale expulsions then.

Whatever the answer to that question—and perhaps it is the wrong question to ask—we need to treat the expulsion from Spain as a separate event, located in its own time and driven by particular contingent pressures. As with England and France we have grown accustomed to see the 1492 expulsion as the natural end to Spain's struggle with its Jewish population. It did not seem natural to those involved. The dynamics of the expulsion suggest that it came as a shock to Jews. They could look back on centuries of life in Spain. Arguably, they were the most highly assimilated and acculturated Jewish community in

Europe. They had survived the violence and conversion in 1391, and as we have seen, continued to thrive in the years of the fifteenth century.

The Spanish monarchy set the expulsion in motion as a way to protect the New Christians from being seduced to return to Judaism (a sign of how powerful Jews must have seemed). The Jewish community offered a reservoir of support and inspiration to disaffected New Christians.[10] Expulsion was a specific royal policy response to a perceived social crisis. (The Spanish could tolerate the presence of Jews under different conditions as they did when they controlled Milan). The language of the expulsion order suggests if anything the continuing vitality of the Jewish community, and attempts in Spain and later in Portugal and Navarre to retain Jewish communities once they converted suggests that they were considered integral parts of society. The expulsions in England and France followed their own trajectory and were triggered by particular historical circumstances. The same is true of Spain in 1492. For the Jews expelled from Spain, the event was profoundly traumatic but could not have been foreseen as the inevitable end of Jewish life in the peninsula.

⁂

The final expulsion of 1492 occurred during a period of European history marked by two movements, the Renaissance and Reformation, that seemed to signal the close of the Middle Ages and the beginning of the long rise of modernity. The expulsion is such a dramatic turning point in the course of Jewish history that it is very seductive to relate it to what appear to be equally dramatic contemporary changes in European society. It has been easy for scholars to see 1492 and its sequelae as a fundamental break in Jewish history because the expulsions occurred around the same time as these two most noticeable "breaks" in

European society. The two phenomena seem to reinforce each other. These two factors, the drama of the expulsions—particularly from Spain in 1492—and the long shadows of periodization cast by the Renaissance and the Reformation, combine to create a sense of a fundamental disjuncture in the experience of Jews in Europe. I think that this emphasis on 1492 and the obsession with the Renaissance and Reformation make it harder to see the continuities in Jewish-Christian relations. I would like to explore here the possibility that relations between Jews and Christians in the fifteenth and sixteenth centuries were characterized by a fundamental continuity from the Middle Ages, one that bridged the apparent divide of the Renaissance and Reformation.

It is only recently that scholars have begun to reconsider whether the Renaissance and the Reformation were truly discrete moments of revolutionary change that marked a "modern" break with a static medieval culture.[11] It may be more useful to think of Renaissances and Reformations in the plural. The terms are useful in describing the heightened awareness of a humanist culture and challenges to church authority.[12] Neither term really defines a period. It is more accurate to imagine an extended late medieval or early modern period that begins in 1348 and ends with the termination of the religious wars in the seventeenth century. The period can be characterized by three trends or movements. The first is the population collapse after the plague and the subsequent late medieval economic depression followed by recovery in the fifteenth century. The second is the displacement of a universal Christian culture of churches, monasteries, and the papacy by national states. The third is the creation of the European empires overseas that created an increasingly global network of production and exchange.[13] Each of these shifts reoriented European society in fundamental ways. But all three took place over generations.

If we embrace this vision of a more gradual transition from medieval to early modern Europe, we have to give up the search for a traumatic division between a medieval and early modern culture and social structure. What remains is a powerful sense of the continuity of basic social structures throughout this period. It is more difficult to argue for a true break in the nature of Jewish-Christian relations when the larger trend in European history is one of fundamental continuity. As I have tried to suggest in the preceding chapters, the experience of Jews in medieval Europe cannot be reduced to a one-dimensional account of persecution and violence. At the same time, by emphasizing that postmedieval social changes created radically new structures for Jewish-Christian relations, historians obscure the survival of essentially medieval conventions of relations. In other words, at least where Jews are concerned, as we move into the fifteenth and sixteenth centuries the structures and nature of Jewish-Christian relations seem remarkably medieval. They reflect a pragmatic tolerance between Christians and Jews and a resilience of Jewish connections to localities. It is the continuity with medieval styles of relations between Christians and Jews that needs to be emphasized and explained.

In the remaining pages of this chapter I hope to suggest that we can see a truer picture of Jewish-Christian relations in Italy and Germany, the countries where Jews continued to live after 1492, by illustrating the essential medievalism in Jewish-Christian relations. Violence against Jews and expulsions of Jews continued to be parts of these worlds. (It is equally true that Jews witnessed generalized societal violence in a world arguably more violent than that of the core medieval centuries.) At the same time, a pragmatic tolerance and stability of relations between Jews and Christians balanced the violence and pressure against Jews. There was no particular turning point or natural resolution of relations between Christians and Jews. What we can find in the early modern period is a continuation of the modus vivendi we have seen in medieval Europe.

The impact of the Reformation and the post-Reformation tolerance helped to shape some of the wide range of reactions to Jews, but I would argue for greater emphasis to be placed on the structures of sixteenth-century society that allowed Jewish communities to survive. The exhaustion of Europe and its belief in theological truths at the end of the religious wars created an opening for a pragmatic vision of skepticism and toleration. This new pragmatism and the associated rise of state involvement in economic life gave Jews great openings to reenter European countries and participate in European economic life.[14] Jonathan Israel is certainly right to point out that European society and economic life was open to Jews in new ways after 1570. But even before the onset of this confessional proto-tolerance, Jews may have benefited from the pragmatic tolerance of the sixteenth century.

Bob Scribner has recently offered a compelling alternative vision of the nature of tolerance in early modern Germany—a vision that seems to be an echo of the pragmatic tolerance we have seen for the Middle Ages as well. This "tolerance of practical rationality" is a vision of society that argues for a widely held acceptance of other people in the small-scale towns and villages of the European countryside. People recognized that there would be differences of opinion among their neighbors since they were increasingly aware that theologians could not come to agreement either.[15]

Echoing other criticisms of Moore's construct, Scribner makes a powerful case that Europe of the sixteenth century (particularly Germany) was not a "persecuting society." Scribner argues that the institutional structures of the sixteenth century could not sustain such persecution over time or great distances. He suggests that popular opinion in the period could not be controlled by bureaucratic elites or "moral crusaders." He cautions us to remember that "within any culture, stereotypes are continually being formed, modified, forgotten, revived, revised and discarded."[16] When we look at the history of Jews during this period, their experiences fit rather well with Scribner's revised sense of a society with a "loose weave" of tolerance. That is, there are moments of intense persecution, when factors such as economic pressures and political or religious rhetoric come together, and other times when these moments passed without producing violence. What is most striking, however, is that Scribner's loosely woven sixteenth century seems to preserve the structures and mentalities that had shaped relations between Jews and Christians in the Middle Ages.

⚘

The experience of Jews in Italy in the early modern period suggests that until the creation of the ghetto in the sixteenth century, there was no real break with the traditions of medieval social relations between Christians and Jews.[17] In many ways, the history of Jews in Italy indicates that instability and resilience characterized their lives throughout the medieval period.

From the fourteenth century, Jews were able to expand their communities, usually centered around networks of banks. In the 15th century communes moved against Jewish lending, and radicalized mendicant preachers stirred up opposition to Jews, resulting in a dramatic contraction of Jewish communities. Jews in southern Italy also endured fluctuating fortunes of expulsions and forced conversions under various rulers, but core populations survived and migrated north.[18] Migration seems to have been an essential part of the Jewish experience in Italy, reaching the levels of what one scholar has even termed nomadism. This does not suggest a lack of roots in any one place, but rather the ability through marriage or business to take up multiple residences with relative ease.[19]

In Rome itself, the experience of Jews mirrors much of the rest of Italy. Jews survived the fourteenth century in Rome quite well. The violence of Cola di Rienzo's coup against the papal state did not set loose any violence or expulsion policies against Jews. The plague, as well, struck Rome just as elsewhere, but Jews did not become victims of fantastical conspiracy accusations.[20] In the years before the Counter-Reformation, the papacy protected Jews in the papal states around Rome.

Jews were able then to navigate through complicated social and political conditions and still remain vital participants in local Italian societies. Despite expulsions and moments of persecution, Italian Jews continually reconstituted their careers and social networks during these late medieval and early modern centuries. What the experience in Italy shows is that there could be a range of reactions to Jews, including violence, judicial persecution, and expulsion, but that ultimately Jews were treated as normal actors in Italian society.

The lives of Jews in the years before the creation of ghettos in the major Italian cities of Venice, Rome, and Florence in the sixteenth century were marked by a high degree of integration into local societies. We now have a brilliant portrait of part of this world in Stefanie Siegmund's study of the creation of the Florentine ghetto.[21] The Medici state created the ghetto, as Siegmund argues, to fulfill the ideology of Counter-Reformation anti-Judaism and to reap the profits of rent generated by the new ghetto. Moreover, the ghetto was an expression of the Medici rulers' desire to regularize and control its diverse population by creating a commune of Jews. When Jews of the Tuscan countryside tried to resist this forced relocation, they called on their neighbors to testify to their honest business practices and good characters. The testimony collected by the Florentine officials, albeit largely ignored, provides eloquent witness to the high degree of acculturation and relatively "normal" social relations between Jews and Christians. Jews in the small towns of the Tuscan countryside were considered long-time residents who had Christian friends, business partners, and employees. The anti-Jewish leaders of the ghetto project could not find a significant number of Christians willing to condemn their Jewish neighbors.

It is unlikely that those kinds of long-term relationships survived the creation of the ghettos where Jews were aggressively removed from the full range of daily social interactions with Christians. However, even with the creation of the ghettos, we may ask if Jews felt profoundly alienated from Italian society. General antagonism toward Jews in the form of ritual murder accusations, attacks, and expulsions declined or disappeared as the ghettos spread.[22] Jews could be tolerated as long as they seemed to be controlled.

We should not romanticize the ghetto, but it did not destroy the Jewish community.[23] (In fact, Siegmund argues that the ghetto actually created the institutions and self-consciousness of the Jewish community). Jews already used to swings in papal and local policy may have been able to take the constraints of the ghetto in stride.[24] One scholar who has studied that world closely concludes:

Jews' perceptions of themselves and their relationship with the outside world in the age of the ghettos were ambivalent and contradictory. On the one hand, a fundamental instability colored ghetto life, owing to the precariousness of existence within the ghetto's walls, marked by overcrowding, conversionary pressure, and growing poverty. On the other hand, Jewish identity was unexpectedly strengthened by the clear-cut delimitation of the community's boundaries. These two contrasting perceptions co-existed, representing two different aspects of the very same situation and at the same time expressing the basic paradox of the ghetto.[25]

The strengthened sense of community and identity may have given Jews the resources to function in the increasingly regulated and religiously aggressive worlds of early modern Italy. Jews who were relocated to ghettos increasingly lived their lives apart from Christians, but in a more constrained way they remained a part of local Italian worlds. They could work and socialize outside the ghetto. The restrained worlds of the ghettos produced the highly acculturated Jews of the early modern world such as come to life in the autobiography of Leone Modena.[26] By the end of the sixteenth century, many aspects of Jewish-Christian relations that had characterized the medieval period in Italy survived into the worlds that had supposedly seen the transformations of Renaissance and Reformation.

Continuity with the medieval past is also seen in Germany where the situation of Jews cannot be reduced to a simple narrative of persecution and expulsion triggered by the end of the Middle Ages. A medieval style of Jewish-Christian interaction, characterized by the predominance of peaceful relations with Christians, resilience of local Jews to outbreaks of violence, and their attachment to specific localities, is evident in the early modern period. Most contacts between Jews and Christians took place in small towns or villages, where the majority of Jews lived.[27] In these small-scale environments, it makes sense that Jews and Christians would have been forced to interact on relatively intimate levels. A recent study of Jewish communities in early modern Germany has collected many references to Jews and Christians forming social bonds at holidays, meals, and weddings, as well as extensive business connections.[28] Jews could, it seems, become burghers or citizens of some of the towns. The communal oaths seemed concerned with tying Jews to a particular town, which reflected the sense that Jews had to be—and could be—made local people.[29] The legislation on restrictions of Jewish dress can be seen, of course, as evidence of that very interaction and acculturation. That interaction went so far that Christians could be found living within Jewish neighborhoods.[30]

These late medieval and early modern German towns were often anxious about the presence of Jews, but this anxiety frequently occurred in the context of concern about people Christians considered outsiders in general. Sometimes Jews were part of that "other" foreign group, but sometimes they were considered local people.[31] When violence did manifest itself, particularly after accusations of host desecration or other alleged Jewish crimes, the episodes were localized and of limited duration.[32] In the cases where the accusations evolved into trials or violence against local communities, we can still see that Jews and Christians lived in a gray zone. Sometimes authorities intervened to stop the proceedings; other times they facilitated the actions against Jews. Sometimes Jews were cruelly punished, other times the animus seems to have spent itself and spared the Jewish community.[33] In Hanau, for example, an initial accusation against a Jew for buying the blood of a murdered Christian boy was stopped when the boy's stepfather withdrew the accusation.[34] Not every Christian in these societies was in thrall to anti-Jewish fantasies.

In the second half of the sixteenth century, Jews appealed successfully on several occasions to imperial courts to have ritual murder accusations quashed. It was at this time that the imperial rulers made a concerted effort to reaffirm their protection of Jews, which was part of their effort to reestablish their position vis à vis the local princes and rulers. Even though the imperial effort may have been motivated by larger political considerations, the blood-libel accusations in the German lands seemed to lose some of their internal coherence. The traditional characteristics of such stories were missing in the narratives at the end of the sixteenth century.[35] Other accusations were aborted for lack of evidence or were clearly attempts to frame local Jews (at least once by a foreign Jew angry at the Frankfurt community).[36]

Even when anti-Jewish feeling resulted in Jews being confined to a particular part of a town, it did not necessarily mean they were cut off from contact with Christians. The displacement of Jews within the towns did not necessarily reflect universally poisoned relations with Christians. In any case, this kind of internal segregation was rarely practiced and had little effect on most Jews. Indeed, many of the towns were so small that enforcing social geography was meaningless. The Jewish population in general was so limited that they were forced to interact with Christians.[37] In general, what defined many of these towns was the continuing contact between Jews and Christians. As one historian has recently concluded: "First of all, we have to say that neither medieval nor post-medieval Jewish culture was the culture of the ghetto." In fact, he continues, "Jewish culture in the *galut* was above all a culture in contact. It accepted new fashions and behaviors and yet continually struggled with the problem of combining non-Jewish cultural achievements with Jewish identity."[38]

To be sure, the Reformation was a factor in the way some Christians behaved toward Jews. Local politics, at least in the towns, seems to have been colored by a new religious consciousness associated with the Reformation. In this sense, the urban expulsions of this period can be said to draw at least some of their energy from the changes associated with the shifting of sacral authority in the Reformation years. The newly reformed citizens of the town began to feel themselves part of a sacred civic group that had to purify its members or at least assert its independence of outside authority. At the same time, Jews became a stalking horse for larger political struggles between different sources of authority, usually that of the town and the nobility.[39] The contested nature of authority in many locations worked in the Jews' favor as a way to counter the animosity that seemed concentrated in the towns.[40] While the Lutheran cities and territories expelled Jews, the ecclesiastical princes of the church, for example, largely tolerated Jews.[41]

It is important to keep in mind, too, that these expulsions are difficult to interpret as reflecting a consistent social policy toward Jews. There is no doubt that expulsions of Jews became in the fifteenth century an increasingly common reaction in Germany. Major towns such as Basel, Cologne, Freiburg, Augsburg, Breslau, Mainz, Bamberg, Ulm, and Nuremburg all expelled Jews in the fifteenth century. Regensburg became the last large city to expel its Jews in 1519, leaving only Frankfurt and Worms as major cities with Jewish populations.[42] Territorial authorities as well expelled Jews. They had to leave Bavaria, the church lands of Mainz, Bamberg and Passau, the duchies of Styria, Carinthia, Krain, Mecklenburg, Brandenburg, Salzburg, and Magdeburg.

Without minimizing the trauma that these decisions inflicted on Jews, it is important to recognize that the expulsions were not uniform or final. They were often temporary. When Jews were expelled from larger towns, they often found refuge in small towns or villages.[43] The lack of a central power made general execution of expulsions difficult. In Cologne and Mainz, the prince-bishop permitted Jews driven from the cities to remain in the countryside or in smaller towns. Shortly after the Jews of Brandenburg were exiled because of a charge of host desecration, its elector proclaimed them innocent and re-admitted them to most of the cities in the electorate. Emperor Maximilian I allowed exiles from Carinthia and Styria to settle elsewhere in his lands. Some rulers tried to make amends for the violence visited on Jews. The Brandenburg elector Joachim II invited the survivors of the Berlin trials (thirty-eight Jews were burned alive) to return and some did. In other cities in the 1500s expulsions could be prevented or the return of Jews sanctioned. When Augsburg decided in 1440 to expel its Jews, they were given two years to leave, hardly conditions of panic or uncontrolled antagonism. Some historians have argued that they merely went to other local towns, and Jews were allowed to conduct business in the town during the day. Certainly, the connection of the Jews to Augsburg was quite strong as some very wealthy Jews offered to pay to remain as citizens.[44] Freder-ick III would later fine the town for improperly expelling the Jews. Even after expulsion, Jews thus often remained attached to their native towns.

Even if we accept that some common Reformation ideology sparked the expulsion efforts of Protestant areas, the actions only succeeded when the constellation of local politics was fa-vorable. This reflects the same kind of contingency that I have tried to argue marks the great expulsions from England, France, and Spain. The experience of Jews in these days was not monolithic. Geography mattered. Jews who lived in the north faced more volatile conditions than those in the south.[45] Despite the constriction and idiosyncratic instability of Jewish life in German lands, I hope it is clearer now that the expulsions (and associated violence or restrictive measures against the Jews) were part of a larger dynamic of relations between Jews and Christians that could oscillate between "acceptance" and "rejection." In this sense, the nature of relations between Jews and Christians reflected a fundamental continuity with the Middle Ages.

In the end, the conflict between Catholicism and Protestantism, while initially radicalizing some Christians, also opened up social space for tolerating Jews. Just as adherents of new denominations had to be tolerated—once it seemed impossible to eliminate them—so too did Jews have to be permitted some space in European society. In this sense, "the coming of multi-confessionalism created, by analogy, more room for the Jews." Christian attention seemed increasingly focused on the radicalized Christian movements such as the Anabaptists, who were seen as more of an immediate threat to the Christian polity.[46] The rise in witchcraft persecution may also have been something of a safety valve for animosities that previously might have been directed against Jews. It is possible, too, that the ongoing threat of the Ottoman Turks distracted Christians from the perceived threats posed by local Jews. (Although the presence of Jews in Ottoman lands may have stimulated anxiety about Jewish conspiracies as it sometimes did in Spain regarding links between Jews in Christian kingdoms and those living under Islam). I wonder, too, if the increasing awareness of the New World and its unconverted peoples presented a new and more important challenge. Surely it was important to convert or at least disperse the Jews, but the great population of newly discovered pagan peoples had to be converted first, since as everyone knew, the Jews would come to Christianity at the end of time.

However, the opening up of toleration associated with confessionalism or a loosely woven pragmatic tolerance could not have succeeded in protecting Jews unless they were already integrated into communities or could draw on traditions that considered them to be part of, even in a marginalized way, European society. It is possible that many people looked to the medieval past for these habits of tolerance. Anna Foa has seen something of this in her recent book where she argues for a persistence of a traditional attitude toward Jews that allowed for a pattern of "alternation of settlement followed by expansion, pogroms, and contraction."[47] The waning of the expulsions of Jews in the sixteenth century and the final reconciliation to a Jewish presence in many European locales may have been a testament to the enduring experience of the Middle Ages and the continuation of those midieval habits during the years of the early modern period.

CONCLUSION

CONVIVENCIA IS TRADITIONALLY used to describe the positive cultural and social interaction between Jews and Christians in medieval Spain. It is easy to romanticize the *convivencia* of Iberia. The reality is more complicated, but historians are reluctant to abandon the term. Recently, scholars have come to an informal consensus on the meaning of *convivencia*. Thomas Glick writes, for example, that "the word, as we use it here, is loosely defined as 'coexistence,' but carries connotations of mutual interpenetration and creative influence, even as it also embraces the phenomenon of mutual friction, rivalry, and suspicion."[1] This produces a "process of normalization of day-to-day interactions, [and] provides the immediate social context for cultural exchange."[2] Finally, Glick argues, "*convivencia*, under any kind of operational definition, must encompass the ability of persons of different ethnic groups to step out of their ethnically bound roles in order to interact on a par with members of competing groups."[3] Benjamin Gampel expands on this definition:

> When we employ the term *convivencia* . . . we are not attempting to conjure up an image of total harmony, of a cosmopolitan setting wherein all faith-communities joyfully infused each other with their particular strengths. Rather we are evoking images of a pluralistic society where communities often lived in the same neighborhoods, engaged in business with each other, and affected and infected each other with their ideas. At the same time, these

groups mistrusted each other and were often jealous of each other's successes, and the ever-present competition among them occasionally turned to hatred.[4]

Convivencia is thought to have existed under Muslim rule in Iberia before the eleventh century. However, even with the success of the *Reconquista* and the flight of Jews away from radicalized Muslim kingdoms to the Christian north, some core elements of *convivencia* survived. This Christian *convivencia* was ultimately more fragile. Gampel, for example, argues that "while the thirteenth century was enlivened by unusually positive interplay among the three faith-communities, we cannot help but notice that with Iberia now the southwesternmost appendage of European civilization, the prejudices of that culture toward other monotheisms began to infiltrate the Christian kingdoms of the peninsula."[5]

Thus, while a precise definition of *convivencia* remains unattainable, scholars seem to think that they know it when they see it. And scholars have definitely not seen it in northern European countries. We have grown used to understanding the medieval period in general as one of intolerance and persecution. Particularly where Jews are concerned, the period after 1096 is treated as one of intensifying persecution and animus toward Jews that led ultimately and inexorably to their expulsion. That sense of Europe as fundamentally hostile to Jews and incapable of any real tolerance animates most of the literature on medieval Jewish-Christian relations.

I hope this study of how Jews lived among Christians has suggested that many of the fundamental characteristics and experiences of *convivencia* can be seen in non-Spanish settings. Jewish-Christian relations in northern Europe is actually *convivencia* in a minor key. Seeing the medieval past in this light will perhaps help to eliminate or at least challenge the false dichotomy between the experience of Jews in Spain (and other Mediterranean settings) and of Jews in northern European societies in the Middle Ages. Jews of England, France, Italy, and Germany were deeply integrated into the rhythms of their local worlds. They faced many of the same challenges and uncertainties as their Christian neighbors. They navigated a world of unexpected violence but recurring stability, ad hoc policies of repression and toleration. All of this suggests that Jewish-Christian relations were dynamic and cannot be understood only in terms of persecution. Jewish-Christian interaction in medieval Europe created if not a history of toleration then habits of tolerance. They were not habits that created pluralism or an anachronistic multiculturalism. Every personal interaction between a Jew and Christian was probably tinged with suspicion and ambivalence on both sides.

Whatever the nature of *convivencia* in Spain and elsewhere in medieval Europe, the expulsions of Jews from western European countries seems to have brought it to an end (but not, of course, in Poland-Lithuania).[6] But simply to say this, too, may be conceding too much to the traditional narrative. Given the alternative reading of the medieval Jewish experience that I have proposed, it is hard for me to accept such an explanation or characterization of the expulsions. The sheer variety, conditions, and duration of the expulsions present a picture of chaos and ad hoc solutions rather than any concerted or consistent cultural or political response. The final period of the expulsions cannot be seen, then, as a fundamental break with the past. The readmissions of the Jews to areas of western Europe, beginning around 1570, is further evidence that social relations between Jews and Christians could reestablish themselves. Jews could not go back to Spain, but over time they tried to reestablish communities in other Western countries where they had lived.

I believe historians, distracted by the Reformation and the putative end of the Middle Ages, have not paid enough attention to the ongoing habits of a pragmatic tolerance that I have tried to describe for the medieval period. Nederman's recent work has demonstrated that such a pragmatic tolerance was a powerful strand in medieval thought, and Scribner has described its survival into the early modern period. For the readmissions to work, whatever new attitude of pragmatism characterized Christian thinking, Jews must have felt some basic level of confidence in their abilities to survive in western European societies, a likely testament to the memory of an effective modus vivendi in these countries. Jews of early modern Europe could look back on centuries of remembered persecution and expulsion, but they also apparently remembered centuries of residence and attachment to their European homes. Coexistence remained a possibility. For that matter, Jews who chose to remain in Poland-Lithuania, and the Christians who welcomed them, did not think it was impossible for Jews and Christians to coexist. Jews and Christians would continue to live together and to live apart. The expulsion of Jews from Iberia—a world of seemingly fewer boundaries for Jews—was final. Ironically, it was the bloody history of medieval Ashkenaz with its recurring persecutions and expulsions—and yet a living habit of tolerance—that provided hope, at least until more modern times, for the future of European Jews.

NOTES

Introduction

1. Jonathan Elukin, "From Jew to Christian? Conversion and Immutability in Medieval Europe," in James Muldoon, ed., *Varieties of Religious Conversion in the Middle Ages* (Gainesville, 1997), 171–90; Elukin, "The Discovery of the Self: Jews and Conversion in the Twelfth Century," in Michael Signer and John Van Engen, eds., *Jews and Christians in Twelfth-Century Europe* (Notre Dame, IN, 2001), 63–77; and Elukin, "Judaism: From Heresy to Pharisee in Medieval Christian Literature," *Traditio* 57 (2002): 49–66.

2. Erich S. Gruen, *Diaspora: Jews amidst Greeks and Romans* (Cambridge, MA, 2002), makes a strong case that Jews had come to terms with their diaspora existence.

3. Yosef Hayim Yerushalmi, Zakhor: *Jewish History & Jewish Memory* (New York, 1989), 44–52.

4. On Jewish historiography, see Ismar Schorsch, *From Text to Context: The Turn to History in Modern Judaism* (Hanover, NH, 1994); Shmuel Feiner, Haskalah *and History: The Emergence of a Modern Jewish Historical Consciousness* (Oxford, 2002); David N. Myers, *Re-Inventing the Jewish Past: European Jewish Intellectuals and the Zionist Return to History* (New York, 1995). My own work on Heinrich Graetz suggests that contemporary concerns often controlled his vision of the Jewish past. See Elukin, "A New Essenism: Heinrich Graetz and Mysticism," *Journal of the History of Ideas* 59 (1998): 135–48.

5. Elukin, "Jacques Basnage and the *History of the Jews:* Polemic and Allegory in the Republic of Letters," *Journal of the History of Ideas* 53 (1992): 603–31. See now Adam Sutcliffe, *Judaism and Enlightenment* (Cambridge, 2003), for an excellent discussion of the limits of Enlightenment thinking about the Jews.

6. Gavin Langmuir, "Majority History and Postbiblical Jews," in his *Toward a Definition of Antisemitism* (Berkeley, 1990), 21–41.

7. Ibid., 35–36.

8. Paul Freedman and Gabrielle M. Spiegel, "Medievalisms Old and New: The Rediscovery of Alterity in North American Medieval Studies," *American Historical Review* 103 (1998): 677–704.

9. Robert I. Moore, *The Formation of a Persecuting Society: Power and Deviance in Western Europe, 950–1250* (Oxford, 1987), and Moore, "Anti-Semitism and the Birth of Europe," in Diana Wood, ed., *Christianity and Judaism, Studies in Church History*, vol. 29 (Oxford, 1992), 33–57. Moore, *The First European Revolution, c. 950–1215* (Oxford, 2000), contains a restatement of the thesis. The interpretations of many recent historians (including Jeremy Cohen, Robert Chazan, Anna Sapir Abulafia, and Gavin Langmuir) echo Moore's approach by trying to identify points when medieval Christian society turned against the Jews.

10. For a sustained critique of Moore, see Cary J. Nederman, "Introduction: Discourses and Contexts of Tolerance in Medieval Europe," in John Christian Laursen and Cary J. Nederman, eds., *Beyond the Persecuting Society: Religious Toleration before the Enlightenment* (Philadelphia, 1998), 13–25. Other works that discuss Moore are Nederman, *Worlds of Difference: European Discourses of Toleration c. 1110–c.1550* (University Park, PA, 2000); Nederman and Laursen, eds., *Difference and Dissent: Theories of Toleration in Medieval and Early Modern Europe* (New York, 1996); and Michael Gervers and James M. Powell, eds., *Tolerance and Intolerance: Social Conflict in the Age of the Crusades* (Syracuse, 2001).

11. See Paul Freedman, " 'The Medieval Other,' The Middle Ages as Other," in Timothy S. Jones and David A. Sprunger, eds., *Marvels, Monsters, and Miracles: Studies in the Medieval and Early Modern Imagination* (Kalamazoo, 2002), 1–27, esp. 4.

12. Ibid., 11.

13. Salo Baron, "Ghetto and Emancipation," *Menorah Journal* 14 (1928): 515–26.

14. Robert E. Lerner, *The Feast of Saint Abraham: Medieval Millenarians and the Jews* (Philadelphia, 2001); John Y. B. Hood, *Aquinas and the Jews* (Philadelphia, 1995); Johannes Heil, " 'Deep Enmity' and/or 'Close Ties?' Jews and Christians before 1096: Sources, Hermeneutics, and Writing History in 1996," *Jewish Studies Quarterly* 9 (2002): 259–306; Ivan Marcus, "A Jewish-Christian Symbiosis: The Culture of Early Ashkenaz," in David Biale, ed., *Cultures of the Jews: A New History* (New York, 2002), 449–518.

15. David Nirenberg, *Communities of Violence: Persecution of Minorities in the Middle Ages* (Princeton, 1996). What I think is equally important to understand, even as we recognize the local and contingent nature of violence, is the impact of that contingent violence on the Jews. It is, I argue below, the Jews' awareness of the transient and contextual nature of the violence against them that encouraged them to remain integrated into their local worlds.

16. Michael André Bernstein, *Foregone Conclusions: Against Apocalyptic History* (Berkeley, 1994).

17. Jeremy Cohen, "The Jews as the Killers of Christ in the Latin Tradition, from St. Augustine to the Friars," *Traditio* 39 (1983): 1–27.

18. Epigraph by Hans Freyer from Herbert Grundmann, *Religious Movements in the Middle Ages,* trans. Steven Rowan (Notre Dame, IN, 1995). I would like to thank one of the readers for Princeton University Press for this quotation.

Chapter One
From Late Antiquity to the Early Middle Ages

1. Scholars have proposed a variety of explanations for this apparent tolerance. See, for example, Solomon Katz, *The Jews in the Visigothic and Frankish Kingdoms of Spain and Gaul* (Cambridge, MA, 1937); Bernard S. Bachrach, *Early Medieval Jewish Policy in Western Europe* (Minneapolis, 1977); Bernard Blumenkranz, *Juifs et Chrétiens dans le monde occidental 430–1096* (Paris, 1960); Mark R. Cohen, *Under Crescent and Cross: The Jews in the Middle Ages* (Princeton, 1994); and Gavin Langmuir, "The Transformation of Anti-Judaism," in *Toward a Definition of Antisemitism*, 63–99.

2. John North, "The Development of Religious Pluralism," in Judith Lieu, John North, and Tessa Rajak, eds., *The Jews Among Pagans and Christians in the Roman Empire* (London, 1992), 174–94. Recent full-scale studies of Jewish life in Late Antiquity can be found in Louis Feldman, *Jew and Gentile in the Ancient World: Attitudes and Interactions from Alexander to Justinian* (Princeton, 1993); Peter Schäfer, *Judeophobia: Attitudes toward the Jews in the Ancient World* (Cambridge, 1997); Seth Schwartz, *Imperialism and Jewish Society, 200 B.C.E. to 640 C.E.* (Princeton, 2001). For the term "micro-Christendom," see Peter Brown, *The Rise of Western Christendom: Triumph and Diversity, A.D. 200–1000* (Cambridge, MA, 1996).

3. Fergus Millar, "The Jews of the Graeco-Roman Diaspora between Paganism and Christianity, AD 312–438," in Lieu et al., *Jews Among Pagans and Christians,* 116–18.

4. See Paula Fredriksen, "*Secundem Carnem:* History and Israel in the Theology of St. Augustine," in William E. Klingshirn and Mark Vessey, eds., *The Limits of Ancient Christianity: Essays on Late Antique Thought and Culture in Honor of R. A. Markus* (Ann Arbor, 1999), 26–42. For the later development of Augustinian tolerance, see Shlomo Simonsohn, *Apostolic See and the Jews: History,* 8 vols. (Toronto, 1991), 7: 6–12. The papal legislation that formalized the church's position, *sicut judeis,* was issued by Calixtus II around 1119 and repeatedly confirmed by popes throughout the Middle Ages. Our first text of *sicut judeis* is from Alexander III's confirmation. For the later history of this ideology, see Walter Pakter, *Medieval Canon Law and the Jews* (Ebelsbach, 1988), and Jeremy Cohen, *Living Letters of the Law: Ideas of the Jew in Medieval Christianity* (Berkeley,1999).

5. Simonsohn, *Apostolic See and the Jews,* 7:7–8.

6. Ramsay MacMullen, *Christianity and Paganism in the Fourth to the Eighth Centuries* (New Haven, 1997). See John Van Engen, "Christening the Romans," *Traditio* 52 (1997): 1–45, on the gradual, and often conflicted process of assuming a Christian identity.

7. Judith Herrin, *The Formation of Christendom* (Princeton, 1987), passim.

8. Panoramic views of the changes during this period can be found in Brown, *Rise of Western Christendom;* Herrin, *Formation of Christendom;* Paul Fouracre, ed., *The New Cambridge Medieval History* (Cambridge, 2000), vol. 1, *c. 500–c.700;* Rosamond McKitterick, ed., *The New Cambridge Medieval History* (Cambridge, 1995), vol. 2, *c. 700–c.900,* and McKitterick, ed., *The Early Middle Ages* (Oxford, 2001), as well as the studies specifically on the integration of non-Roman peoples into the empire: Walter Pohl and Helmut Reimitz, eds., *Strategies of Distinction: The Construction of Ethnic Communities, 300–800* (Leiden, 1998); Leslie Webster and Michelle Brown, eds., *The Transformation of the Roman World, AD 400–900* (Berkeley, 1997), especially the essay by Walter Pohl, "The Barbarian Successor States," 33–47; Pohl, ed., *Kingdoms of the Empire: The Integration of Barbarians in Late Antiquity* (Leiden, 1997); G. Ausenda, ed., *After Empire: Towards an Ethnology of Europe's Barbarians* (Suffolk, UK, 1995), especially, Ian Wood, "Pagan Religion and Superstitions East of the Rhine from the Fifth to the Ninth Century," 253–79; and see Patrick Amory's discussion in *People and Identity in Ostrogothic Italy, 489–554* (Cambridge, 1997), 13–42. I have not been able to take

advantage of several new books on the fall of Rome and the creation of medieval society by Chris Wickham, Bryan Ward-Perkins, and Julia Smith.

9. Scott Bradbury, ed. and trans., *Severus of Minorca, Letter on the Conversion of the Jews* (Oxford, 1996). See Bradbury's excellent introduction, and as well, E. D. Hunt, "St. Stephen in Minorca: An Episode in Jewish-Christian Relations in the Early 5th Century AD," *Journal of Theological Studies* n.s. 33 (1982): 106–23; Norman Roth, "Bishops and Jews in the Middle Ages," *Catholic Historical Review* 80 (1994): 1–17; Carlo Ginzburg, "The Conversion of the Minorcan Jews (417–18): An Experiment in History of Historiography," in Scott L. Waugh and Peter D. Diehl, eds., *Christendom and its Discontents: Exclusion, Persecution, and Rebellion, 1000–1500* (Cambridge, 1996), 207–19.

10. Bradbury, *Severus*, chapter 6, 85.

11. Ibid., 5, 85.

12. Ibid., 13, 93.

13. Ibid., 25.

14. Peter Brown, *The Cult of the Saints: Its Rise and Function in Latin Christianity* (Chicago, 1981). See now, James Howard-Johnston and Paul Antony Hayward, eds., *The Cult of Saints in Late Antiquity and the Middle Ages: Essays on the Contribution of Peter Brown* (Oxford, 1999).

15. Bradbury, *Severus*, 16, 97.

16. Ibid., 18, 101. On violence by Jews against Christians, particularly in alliance with pagans or opponents of Catholicism, see Ibid., 56.

17. Ibid., 8, 87.

18. Ibid., 13, 93.

19. Ibid., 71.

20. Ibid., 70.

21. Ibid., 26, 119–20.

22. Recent work on Gregory that tries to explain the form and goals of the *History* include Martin Heinzelmann, *Gregory of Tours: History and Society in the Sixth Century*, trans., Christopher Carroll (Cambridge, 2001); Walter Goffart, *The Narrators of Barbarian History (AD 550–800): Jordanes, Gregory of Tours, Bede and Paul the Deacon* (Princeton, 1988); Giselle de Nie, *Views from a Many-windowed Tower: Studies in the Imagination in the Works of Gregory of Tours* (Amsterdam, 1987); and Kathleen Mitchell and Ian N. Wood, eds., *The World of Gregory of Tours* (Leiden, 2002).

23. Gregory of Tours, *The History of the Franks*, trans., Lewis Thorpe (Harmondsworth, 1974).

24. Ibid., book 9, chapter 4, 483.

25. Ibid., 10, 13, 560.

26. See the insightful caution by Yitzhak Hen that texts actually composed under the Merovingians—rather than later anti-Merovingian Carolingian texts recounting the earlier period—contain very little evidence of real paganism. "Paganism and Superstitions in the Time of Gregory of Tours: Une Question mal posée!" in Mitchell and Wood, eds., *World of Gregory of Tours*, 229–40. See as well, Hen, *Culture and Religion in Merovingian Gaul, A.D. 481–751* (Leiden, 1995).

27. Gregory of Tours, *Glory of the Martyrs*, trans., Raymond Van Dam (Liverpool, 1988), 29–30.

28. Blumenkranz, *Juifs et Chrétiens*, 12–33.

29. Gregory, *History*, 5, 6, 263–64.

30. Ibid., 8, 1, 433.

31. Ibid., 6, 5, 330.

32. Ibid., 6, 5, 330.

33. Ibid., 6, 5, 330.

34. Avril Keely, "Arians and Jews in the Histories of Gregory of Tours," *Journal of Medieval History* 23 (1997): 103–15.

35. Gregory, *History*, 6, 6, 334.

36. Translated in Raymond Van Dam, *Saints and Their Miracles in Late Antique Gaul* (Princeton, 1993), 278–79.

37. Gregory, *History*, 7, 23, 405–6.

38. Ibid., 7, 23, 406.

39. Ibid., 7, 23, 406.

40. Ibid., 7, 23, 406.

41. Ibid., 6, 17, 347–48.

42. Ibid., 6, 17, 348.

43. Ibid., 6, 17, 348.

44. Ibid., 6, 17, 348.

45. See Peter Brown's characterization of this violence in "Gregory of Tours: Introduction," in Mitchell and Wood, eds., *World of Gregory of Tours*, 11.

46. The Jews had lived in Clermont for more than a hundred years or more. They had been on good terms with Sidonius Appolinaris, the bishop of Clermont, in the later fifth century. See Bachrach, *Early Medieval Jewish Policy*, 160, n. 26 for Sidonius's letters regarding the Jews.

47. Gregory, *History*, 5, 11, 266. We know of the violence from a parallel account by the poet Venantius Fortunatus. See Brian Brennan, "The Conver-

sion of the Jews of Clermont in AD 576," *Journal of Theological Studies* 36 (1985): 321–37.

48. Gregory, *History*, 5, 11, 267. King Dabobert I (622–38) apparently demanded the Jews convert or be exiled in the 630s. J. M. Wallace-Hadrill, trans., *Chronicle of Fredegar* (London, 1960), 53–54.

49. Amnon Linder, ed. and trans., *Jews in the Legal Sources of the Early Middle Ages* (Detroit, 1997), 475–76. Mâcon (581–83).

50. Blumenkranz, *Juifs et Chrétiens*, 35–37. See Herwig Wolfram, *History of the Goths* (Berkeley, 1988), 23, citing Procopius that the Jews of Naples supported the Ostrogoths.

51. Blumenkranz, *Juifs et Chrétiens*, 6–10, and Katz, *Jews in the Visigothic and Frankish Kingdoms*, 163–35.

52. Salo Baron, *A Social and Religious History of the Jews*, 2nd ed., 18 vols. (New York, 1952–), 3:48.

53. See Walter Pohl, "Telling the Difference: Signs of Ethnic Identity," in Pohl and Reimitz, eds., *Strategies of Distinction*, 67.

54. Simon Goldhill, *The Temple of Jerusalem* (Cambridge, MA, 2005), 103.

55. For an overview of early Italian history, see Chris Wickham, *Early Medieval Italy: Central Power and Local Society 400–1000* (Totowa, NJ, 1981).

56. Gregory the Great, *Saint Gregory the Great Dialogues*, trans., Odo John Zimmerman, O.S.B., *Fathers of the Church*, 106 vols. (New York, 1959), 39:74.

57. On Gregory, see Carole Straw, *Gregory the Great: Perfection in Imperfection* (Berkeley, 1988); Robert A. Markus, *Gregory the Great and His World* (Cambridge, 1997); and John C. Cavadini, ed., *Gregory the Great: A Symposium* (Notre Dame, IN, 1995).

58. Gregory, *Dialogues*, 39:121–22.

59. Linder, *Legal Sources*, 423. Gregory to Peter the Subdeacon (July 592). See Robert Markus, "The Jew as a Hermeneutic Device: The Inner Life of a Gregorian *Topos*," in Cavadini, ed., *Gregory the Great: A Symposium*, 1–15.

60. Linder, *Legal Sources*, 417. Gregory to Peter, Bishop of Terracina (March 591)

61. Ibid., 419. Gregory to Vergilius and to Theodore, Bishop of Massilia in Gaul (June 591).

62. Ibid., 433–34. Gregory to Victor, Bishop of Palermo (June 598).

63. Ibid., 434–35. Gregory to Fantinus the Protector (October 598).

64. Ibid., 424. Gregory to Libertinus Praetor of Sicily (May 593).

65. Ibid., 416–17. Jews experimenting with their own religious identity was not a new problem. At the end of the fifth century, Pope Gelasius had

written about one Jew: "The Most Illustrious Telesinus (although he seems to be of the Jewish credulity he strives so hard to find favor with us that we should justly be obliged to call him one of ours), requested us particularly in favor of his relative Antonius, that we should have to recommend him to your Affection."

66. Ibid., 420, Gregory to Peter the Subdeacon (August 591).

67. Ibid., 438 (July 599).

68. Ibid., 432–33, Gregory to the Protector Fantinus (c. 594).

69. Ibid., 428, Gregory to Anthemius the Subdeacon (July 594).

70. Chris Wickham, "Society," in McKitterick, ed., *Early Middle Ages*, 84.

71. Linder, *Legal Sources*, 440. Gregory to Brunhild Queen of the Franks. "We are utterly amazed that in your kingdom you permit the Jews to possess Christian slaves. For what are all Christians but members of Christ. . . . We request, therefore, that a constitution of your Excellence should remove this wicked evil from her kingdom."

72. Ibid., 436–37, Gregory to Fortunatus Bishop of Naples (February 599). An earlier letter from Pope Gelasius (492–96) contained a careful instruction to check on the story of a Christian slave who claimed to have been circumcised by his Jewish owner and was now seeking freedom after having fled to a church. It seemed that the rights of any slave owner, even a Jewish one, had to be protected. Ibid., 416–17.

73. Ibid., 426–27. Gregory to Venantius Bishop of Luni (May 594). See discussion in Walter Pakter, "Early Western Church Law and the Jews," in Harold W. Attridge and Gohei Hata, eds., *Eusebius, Christianity, and Judaism* (Detroit, 1992), 718–19.

74. Linder, *Legal Sources*, 426–27.

75. P. D. King, *Law and Society in the Visigothic Kingdom* (Cambridge, 1972), 132. On the Visigoths see E. A. Thompson, *The Goths in Spain* (Oxford, 1969); Norman Roth, *Jews, Visigoths, and Muslims in Medieval Spain* (Leiden, 1994); Edward James, ed., *Visigothic Spain: New Approaches* (Oxford, 1980); Alberto Ferreiro, ed., *The Visigoths: Studies in Culture and Society* (Leiden, 1999), especially R. Gonzalez-Salinero, "Catholic Anti-Judaism in Visigothic Spain," 123–51; Peter Heather, ed., *The Visigoths From the Migration Period to the Seventh Century: An Ethnographic Perspective* (Woodbridge, Suffolk, UK, 1999), especially Heather's article "The Creation of the Visigoths," 43–92; and Rachel Stocking, *Bishops, Councils, and Consensus in the Visigothic Kingdom, 589–633* (Ann Arbor, 2000). For an interesting discussion of Visigothic attitudes towards the Jews, see Jeremy duQuesnay Adams, "Ide-

ology and the Requirements of 'Citizenship' in Visigothic Spain: The Case of the *Judaei,*" *Societas* 2 (1972): 317–32.

76. King, *Law and Society,* 145–51; J. N. Hillgarth, "Popular Religion in Visigothic Spain," in James, ed., *Visigothic Spain,* 3–61.

77. On this, see Pablo Diaz and Ma R. Valverde, "The Theoretical Strength and Practical Weakness of the Visigothic Monarchy of Toledo," in Frans Theuws and Janet L. Nelson, eds., *Rituals of Power: From Late Antiquity to the Early Middle Ages* (Leiden, 2000), 82–83.

78. *Lives of the Visigothic Fathers,* ed. and trans., A. T. Fear (Liverpool, 1997), 74–75.

79. King, *Law and Society,* 131.

80. Ian N. Wood, "Social Relations in the Visigothic Kingdom from the Fifth to the Seventh Century: The Example of Merida," in Heather, ed., *Visigoths: From the Migration Period to the Seventh Century,* 205.

81. Katz, *Jews in the Visigothic and Frankish Kingdoms,* 11.

82. Ibid., 11; Bachrach, *Early Medieval Jewish Policy,* 7.

83. Katz, *Jews in the Visigothic and Frankish Kingdoms,* 11–12.

84. B. Albert, "Isidore of Seville: His Attitude towards Judaism and His Impact on Early Canon Law," *Jewish Quarterly Review* 80 (1990): 207–20.

85. Bachrach, *Early Medieval Jewish Policy,* 10.

86. Katz, *Jews in the Visigothic and Frankish Kingdoms,* 13.

87. Ibid., 13.

88. Ibid., 14.

89. Ibid., 15.

90. Ibid., 15–16.

91. Ibid., 16.

92. Ibid., 17; and Bachrach, *Early Medieval Jewish Policy,* 18–19.

93. Katz, *Jews in the Visigothic and Frankish Kingdoms,* 17–19.

94. Ibid., 20–21.

95. Ibid., 21.

96. King, *Law and Society,* 22.

CHAPTER TWO
FROM THE CAROLINGIANS TO THE TWELFTH CENTURY

1. For surveys of Carolingian history, see McKitterick, *Early Middle Ages 400–1000,* and the second volume of *The New Cambridge Medieval History c. 700–c. 900.*

2. Pseudo-Jerome, *Quaestiones on the Book of Samuel*, ed. A. Saltman (Leiden, 1975), 7–8.

3. Bat-Sheva Albert, "Adversus Iudaeos in the Carolingian Empire," in Ora Limor and Guy G. Stroumsa, eds., *Contra Iudaeos: Ancient and Medieval Polemics between Christians and Jews* (Tübingen, 1996), 120.

4. Heinz Schreckenberg, "Die patristiche Adversus-Judaeos Thematik im Spiegel der karolingischen Kunst," *Bijdragen, tijdschrift voor filosofie en theologie* 49 (1988): 119–38. See Johannes Heil, "Labourers in the Lord's Quarry: Carolingian Exegetes, Patristic Authority, and Theological Innovation, a Case Study in the Representation of Jews in Commentaries on Paul," in Celia Chazelle and Burton Van Name Edwards, eds., *The Study of the Bible in the Carolingian Era* (Turnhout, 2003), 75–96. Heil reinforces the idea that most Carolingian exegetes treated Jews as a theological construct—the means to indict Christian sin—rather than as a real social problem. Most echoed the Augustinian sensibility that the Jews would be saved at the end of time. Heil provides several interesting examples of exegetes who seemed to voice harsher judgments against Jews, including a reticence about their future salvation. He sees the roots of these more intolerant opinions in the commentators' engagement with the debate over predestination. Some advocates of a strict predestination saw the Jews' damnation as the sign of God's ability to choose the elect.

5. Mayke de Jonge, "Old Law and New-Found Power: Hrabanus Maurus and the Old Testament," in J. W. Drijvers and A. A. MacDonald, eds., *Centres of Learning: Learning and Location in Pre-Modern Europe and the Near East* (Leiden, 1995), 161–76; De Jonge, "Exegesis for an Empress," in Esther Cohen and De Jonge, eds., *Medieval Transformations: Texts, Power, and Gifts in Context* (Leiden, 2001), 69–101; Mary Garrison, "The Franks as the New Israel? Education for an Identity from Pippin to Charlemagne," in Yitzhak Hen and Matthew Innes, eds., *The Uses of the Past in the Early Middle Ages* (Cambridge, 2000), 114–62. And in the same volume, De Jonge, "The Empire as *Ecclesia*: Hrabanus Maurus and Biblical *Historia* for Rulers," 191–227.

6. Elukin, "Judaism: From Heresy to Pharisee in Christian Literature," 63.

7. See Janet Nelson, "On the Limits of the Carolingian Renaissance," in Janet Nelson, ed., *Politics and Ritual in Early Medieval Europe* (London, 1986), 63–67. See the discussion of religion by Julia Smith in McKitterick, ed., *The New Cambridge Medieval History*, 2:654–81; and the chapter by De Jonge in McKitterick, ed., *Early Middle Ages*, 131–61. See Celia Chazelle, *The Crucified God in the Carolingian Era: Theology and Art of Christ's Passion*

(Cambridge, 2001) for a discussion of the centrality of Jesus' incarnation in some elite sectors of Carolingian religious culture.

8. Bachrach, *Early Medieval Jewish Policy*, 70.

9. Michael Toch, "Jews and Commerce: Modern Fancies and Medieval Realities," in. S. Cavaciocchi, ed., *Il Ruolo Economico delle Minoranze in Europa. Secc. XIII–XVIII* (Florence, 2000), 43–58. See also his "Economic Activities of German Jews in the 10th to 12th Centuries: Between Historiography and History," in Yom Tov Assis, Jeremy Cohen, Aharon Kedar, Ora Limor, and Michael Toch, eds., *Facing the Cross: The Persecutions of 1096 in History and Historiography* (Jerusalem, 2000), 32–55 [Hebrew]. On the centrality of the slave trade to the early medieval economy, see Michael McCormick, *Origins of the European Economy: Communications and Commerce, A.D. 300–900* (Cambridge, 2001), 244–54 and 733–99. Jews themselves were not always safe from slavery. Bodo the deacon, the famous convert to Judaism in the ninth century, had made his way to Spain initially with Jews whom he had brought along to sell. *The Annals of St-Bertin*, trans. Janet L. Nelson (Manchester, 1991), 43.

10. Linder, *Legal Sources*, 444–45. Stephen the pope to Aribert, archbishop of Narbonne.

11. We have only the problematic text of several Carolingian laws, grouped under the title *Capitula de Iudaeis*, which purports to regulate Jewish involvement in mints, limitations on loans, labor of Christians on Sunday, and storage of goods outside the marketplace. Bachrach, *Early Medieval Jewish Policy*, 169, n.49.

12. Ibid., 92–93. Louis the Pious may have issued more general laws concerning the Jews, but his active interventions were on a case-by-case basis.

13. Historians have echoed this conclusion. See, for example, Cohen, *Under Crescent and Cross*, 50. For further discussion of Jews as merchants, see Katz, *Jews in the Visigothic and Frankish Kingdoms*, 124–38. Bachrach, *Early Medieval Jewish Policy*, 166, n.25, emphasizes the important role of Jews in long-distance trade.

14. Nelson, *Annals of St-Bertin*, 41–42.

15. Ibid., 65, for Bordeaux, and 74, for Barcelona.

16. Ibid., 202.

17. Indeed, Agobard's familiarity with Jewish customs and laws suggest relatively accurate knowledge of and perhaps access to the contemporary Jewish population in Lyon. R. Bonfil, "The Cultural and Religious Traditions of French Jewry in the Ninth Century, as Reflected in the Writings of Agobard of Lyons," in Joseph Dan and Joseph Hacker, eds., *Studies in Jewish Mysticism,*

Philosophy, and Ethical Literature Presented to Isaiah Tishby on his Seventy-fifth Birthday (Jerusalem, 1986), 327–49 [Hebrew].

18. Heil, "Labourers in the Lord's Quarry," 91.

19. Amolo, *Epistola seu Liber contra Judaeos,* cited in Albert, "Adversus Judeos," 138, n.106. One historian has recently suggested that these Jewish communities were able to grow by converting local people. R. H. Bautier, "L'Origine des population juives de la France médiévale," in X. Barral et al., eds., *Catalunya I França Meridional a L'Entorn de L'Ang Mil* (Barcelona, 1991), 302–16. Cited in McCormick, *Origins,* 793, n.10.

20. McCormick, *Origins,* 675–77.

21. Ibid., 677.

22. Ibid., 667.

23. Cited in Katz, *Jews in the Visigothic and Frankish Kingdoms,* 133.

24. See Elisheva Carlebach, "The Sabbatian Posture of German Jewry," *Jerusalem Studies in Jewish Thought* 16–17 (2001): 1–29, where she argues that scholars have overlooked or dismissed the seriousness of messianic movements among European Jews in the Middle Ages. The emigration of rabbis to Israel as an expression of that messianism may represent a popular movement of some depth. In any event, it seems safe to say that migration to Palestine was not a permanent feature of European Jewry and when it did occur, seemed to be at moments of extreme religious enthusiasm.

25. On Lyons, see Florus of Lyons, *Epistola,* cited in Albert, "Adversus Judeos," 136, n.95. On the Venetians, see McCormick, *Origins,* 796.

26. Seth Schwartz makes a similar argument about the impact of Roman imperialism for the formative period of rabbinic Judaism in *Imperialism* and *Jewish Society, 200 B.C.E. to 640 C.E.*

27. On slavery, see the discussion in David A. E. Pelteret, *Slavery in Early Mediaeval England: From the Reign of Alfred until the Twelfth Century* (Woodbridge, Suffolk, UK, 1995), 1–24. The literature on the feudal transformation is extensive. One place to begin are the contributions by Pierre Bonassie, Dominique Barthélemy, and Elizabeth A. R. Brown in Lester K. Little and Barbara H. Rosenwein, eds., *Debating the Middle Ages: Issues and Readings* (Oxford, 1998), 105–69.

28. Michael Toch, "The Formation of a Diaspora: The Settlement of Jews in the Medieval German *Reich,*" *Aschkenas* 7 (1997): 55–79. See Dean Philip Bell, *Sacred Communities: Jewish and Christian Identities in Fifteenth-Century Germany* (Leiden, 2001), 126–48; and Bell, "Jews," in John M. Jeep, ed., *Medieval Germany: An Encyclopedia* (New York, 2001), 411.

29. Blumenkranz, *Juifs et Chrétiens*, 24.

30. See Robert Chazan, *European Jewry and the First Crusade* (Berkeley, 1987), 31–33.

31. Ibid., 30.

32. On attacks on Jews in Italy, see Kenneth R. Stow, *Alienated Minority: The Jews of Medieval Latin Europe* (Cambridge, MA, 1992), 68.

33. Ibid., 79–80.

34. *The Chronicle of Ahimaaz*, ed. and trans., Marcus Salzman, (New York, 1966), 61.

35. Ibid., 62.

36. Ibid., 66.

37. Ibid., 67.

38. Ibid., 84. Stephen D. Benin, "Jews, Muslims and Christians in Byzantine Italy," in Benjamin H. Hary, John L. Hayes, Fred Astren, eds., *Judaism and Islam: Boundaries, Communication and Interaction. Essays in Honor of William M. Brinner* (Leiden, 2000), 27–35.

39. David Warner, trans., *Ottonian Germany: the Chronicon of Thietmar of Merseburg* (Manchester, 2000), 144. For a survey of Jewish experience in medieval Germany, see Michael Toch, *Die Juden im Mittelalterlichen Reich* (Munich, 1998), with its very extensive bibliography. And now, see Alfred Haverkamp, ed., *Geschichte der Juden von der Nordsee bis zu den Sudalpen. Kommentiertes Kartenwerk*, 3 vols. (Hanover, 2002), and other works in Haverkamp's series, Forschungen zur Geschichte der Juden, including Christoph Cluse, *Studien zur Geschichte der Juden in dem mitteralterichen Niederlanden* (Hanover, 2000).

40. Johannes Heil, " 'Deep enmity' and/or 'Close ties?' Jews and Christians before 1096," 302. Heil's article is an excellent discussion of the problems of using sources designed for overtly religious purposes as evidence of the mentality of people in their everyday lives.

41. Ibid., 302.

42. Warner, *Chronicon of Thietmar of Merseburg*, 286.

43. Blumenkranz, *Juifs et Chrétiens*, 42–43.

44. Haverkamp, *Medieval Germany 1056–1273* (Oxford, 1988), 163, and 212–15. Henry IV later granted charters of protection to the Jewish communities in Speyer and Worms.

45. Chazan, *Church, State, and Jew in the Middle Ages* (New York, 1980), 58.

46. Ibid., 58. Heil, " 'Deep enmity,' " 278: "Furthermore, Bishop Rudiger's concerns about the 'animosity of the people' fits in with his more far-reaching

and reasonable concerns about the general 'disturbance of the kingdom,' as he puts it, at that time. The situation in the years of the German *Investitur-streit* was more than critical, and not only for Jews: the effects of the civil war, the three-sided struggle waged by the partisans of Pope, king and princes, the repeated excommunication of King Henry IV, the election of a rival king, would leave traces in every act that the prominent bishop and staunch supporter of King Henry undertook in those years."

47. Ibid., 58

48. Ibid., 58

49. Ibid., 58.

50. Shlomo Eidelberg, *The Jews and the Crusaders: The Hebrew Chronicles of the First and Second Crusades* (Madison, 1977), 71.

51. Ibid., 71.

52. Ibid., 71.

53. Ibid., 71. This is an epilogue added to the Solomon Bar Simson chronicle. As Heil points out, the Hebrew account may be romanticizing the tragic nature of the history in Mainz as a large Jewish community continued to exist in Mainz after 1084. Heil, " 'Deep enmity,' " 278: "For that reason, we conclude that even assuming that the Jews of Mainz faced actual threats around 1084, only a part of the community was so alarmed that they decided to leave for Speyer."

54. Eidelberg, *Jews and Crusaders,* 71.

55. Chazan, *Church, State, and Jew,* 60–63.

56. Ibid., 60.

57. Ibid., 61.

CHAPTER THREE

CULTURAL INTEGRATION IN THE HIGH MIDDLE AGES

1. Robert Bartlett, *The Making of Europe: Conquest, Colonization, and Cultural Change, 950–1350* (Princeton, 1993), provides the best recent survey of the creation of an increasingly homogenous Christian culture in the Middle Ages.

2. Giles Constable, *The Reformation of the Twelfth Century* (Cambridge, 1996), 5.

3. Caroline Walker Bynum, "Did the Twelfth Century Discover the Individual?" in her *Jesus as Mother: Studies in the Spirituality of the High Middle Ages* (Berkeley, 1982), 85.

152

4. Ivan Marcus, "The Dynamics of Jewish Renaissance and Renewal in the Twelfth Century" in Signer and Van Engen, eds., *Jews and Christians in Twelfth-Century Europe*, 39.

5. Ibid., 33. Jeremy Cohen has recently posited the same kind of cultural commonality for the issue of doubt; see "Between Martyrdom and Apostasy: Doubt and Self-Definition in Twelfth-Century Ashkenaz," *Journal of Medieval and Early Modern Studies* 29 (1999): 463: "Whether or not such uncertainty succeeded in uprooting a Jew from the Jewish community, it bespoke his or her participation in a medieval European experience that included both Christians and Jews. The high Middle Ages comprised a period of doubt, dissent, and experimentation, just as they fostered growth and conquest in all realms of European life. "

6. Shmuel Shepkaru, "To Die for God: Martyrs' Heaven in Hebrew and Latin Crusade Narratives," *Speculum* 77 (2002): 311–41. Kenneth Stow, "Conversion, Apostasy, and Apprehensiveness: Emicho of Flonheim and the Fear of Jews in the Twelfth Century," *Speculum* 76 (2001): 911–33.

7. Chazan, *European Jewry*, 132–36.

8. Marcus, "A Jewish-Christian Symbiosis," 467–68.

9. Ivan Marcus, *Rituals of Childhood: Jewish Acculturation in Medieval Europe* (New Haven, 1996). Other studies include Arthur Green, "*Shekhinah*, the Virgin Mary, and the *Song of Songs*: Reflections on a Kabbalistic Symbol in Its Historical Context," *AJS Review* 26 (2002): 1–52; Yisrael Yuval, "Easter and Passover As Early Jewish-Christian Dialogue," in Paul F. Bradshaw and Lawrence A. Hoffman, eds., *Passover and Easter: Origin and History to Modern Times* (South Bend, Ind., 1999), 98–124, and in the same collection, Yuval, "Passover in the Middle Ages," 127–60; Yuval, "Vengeance and Damnation, Blood and Defamation," *Zion* 58 (1993): 33–90 [Hebrew], and the responses in *Zion* 59 (1994); Yuval, " 'They Tell Lies: You Ate the Man': Jewish Reactions to Ritual Murder Accusations," in Anna Sapir Abulafia, ed., *Religious Violence between Christians and Jews: Medieval Roots, Modern Perspectives* (New York, 2002), 86–106; and Yuval, "Jews and Christians in the Middle Ages: Shared Myths, Common Language," in Robert. S. Wistrich, ed., *Demonizing the Other: Antisemitism, Racism, and Xenophobia* (Singapore, 1999), 88–107.

10. Joseph Shatzmiller, "Jews, Pilgrimage, and the Christian Cult of Saints," in Alexander Callander Murray, ed., *After Rome's Fall: Narrators and Sources of Early Medieval History. Essays presented to Walter Goffart* (Toronto, 1998), 337–48, see especially 342–43: "As if to counter such challenges of the

Christian cult of saints and relics, Jews boasted about miracles that occurred in the framework of their own religion."

11. Beryl Smalley, *The Study of the Bible in the Middle Ages* (Oxford, 1983). William Jordan, *French Monarchy and the Jews: From Philip Augustus to the Last Capetians* (Philadelphia, 1989), 11–15, for the difficulties and lack of mutual understanding involved in such a dialogue. Gilbert Dahan, *Les Intellectuals Chrétiens et les Juifs au Moyen Age* (Paris, 1990). See Hyam Maccoby, *Judaism on Trial: Jewish-Christian Disputations in the Middle Ages* (Rutherford, NJ, 1982).

12. On disputations and polemical literature, see the entry on disputations in Norman Roth, ed., *Medieval Jewish Civilization: An Encyclopedia* (New York, 2003), 212–18. See also Hanne Trautner-Kromann, *Shield and Sword: Jewish Polemics against Christianity and the Christians in France and Spain from 1100–1500* (Tübingen, 1993). Marc Michael Epstein, *Dreams of Subversion in Medieval Jewish Art and Literature* (University Park, PA, 1997), 114: "A strong sense of exile and persecution, but at the same time, a faith in the inevitability of the survival, and indeed of the triumph of the Lions of Judah pervades medieval Jewish culture."

13. Miri Rubin, *Gentile Tales: The Narrative Assault on Late Medieval Jews* (New Haven, 1999), 35, although as Rubin notes there was an increasing level of violence against the Jews in these tales after the thirteenth century.

14. Robert Stacey, "The Conversion of Jews to Christianity in Thirteenth-Century England," *Speculum* 67 (1992): 269.

15. Ibid., 274–76.

16. Alfred Haverkamp, "Baptised Jews in German Lands during the Twelfth Century," in Signer and Van Engen, eds., *Jews and Christians,* 265 and 274–75, for other accounts of highly placed Jewish converts. And Haverkamp, *Medieval Germany,* 215. Joshua had a reputation for going about in clothes more suitable for a knight. He converted only after long discussions and demands by the archbishop.

17. Haverkamp, "Baptised Jews," 267.

18. Ibid., 271–72.

19. Elukin, "From Jew to Christian? Conversion and Immutability in Medieval Europe," 171–90.

20. See, for example, Richard Southern, *The Making of the Middle Ages* (New Haven, 1959) 219–57; and Colin Morris, *The Discovery of the Individual: 1050–1200* (New York, 1973).

21. Elukin, "The Discovery of the Self: Jews and Conversion in the Twelfth Century," 63–77. The following discussion is largely drawn from this article.

22. Innocent III wrote to Peter of Corbeil, archbishop of Sens on behalf of a convert formerly named Isaac. Peter was apparently reluctant to help support the new convert and his family, so Innocent tried to make the former Jews more attractive by relating the story of their conversion, which involved the multiplication of the host hidden in the family's home by a Christian servant. Innocent was trying to give the Jews a pedigree or seal of authenticity to their conversion. See Shlomo Simonsohn, *The Apostolic See and the Jews*, 1:98, no. 93 (Rome, June 8, 1213).

23. Paul J. Archambault, trans., *A Monk's Confession: The Memoirs of Guibert of Nogent* (University Park, PA, 1996), 112. For a recent discussion of Guibert, see Jay Rubenstein, *Guibert of Nogent: Portrait of a Medieval Mind* (New York, 2002).

24. *Monk's Confession*, 112.

25. Ibid., 112.

26. Ibid., 113.

27. See Karl F. Morrison's translation in his *Conversion and Text: The Cases of Augustine of Hippo, Herman-Judah, and Constantine Tsatsos* (Charlottesville, 1992). For a discussion of the contested authenticity of the text, see Jean-Claude Schmitt, *La conversion d'Hermann le juif: autobiographie, histoire et fiction* (Paris, 2003).

28. Morrison, *Conversion and Text*, 86.

29. Ibid., 86.

30. Ibid., 101.

31. Ibid., 101.

32. Sander L. Gilman, *Jewish Self-Hatred* (Baltimore, 1986), 31.

33. See Anna Sapir Abulafia's discussion, *Christians and Jews in the Twelfth-Century Renaissance* (London, 1995).

CHAPTER FOUR
SOCIAL INTEGRATION

1. Chazan, "The Anti-Jewish Violence of 1096: Perpetrators and Dynamics," in Abulafia, ed., *Religious Violence between Christians and Jews*, 22. Chazan, *European Jewry*, 40–9.

2. Chazan, *European Jewry*, 227, 245, 289.

3. Ibid., 246

4. Ibid., 246.

5. Ibid., 251.

6. Ibid., 246.

7. Ibid., 253.

8. Ibid., 268–69.

9. Ibid., 268–69.

10. Ibid., 227.

11. Ibid., 289–90. See also on 284 the account of the mayor of Moers who tried at first to intercede for the Jews.

12. Ibid., 291.

13. Haverkamp, *Medieval Germany*, 122–23. The emperor could take revenge after the massacres as he did against Archbishop Ruthard of Mainz.

14. Chazan, *European Jewry*, 274.

15. Ibid., 274

16. Ibid., 288.

17. Ibid., 249.

18. Ibid., 231–32.

19. Ibid., 261.

20. Ibid., 282.

21. See Yerushalmi, "Exile and Expulsion in Jewish History," in Benjamin Gampel, ed., *Crisis and Creativity in the Sephardic World: 1391–1648* (New York, 1997), 3–23, for a penetrating discussion of how Jews gave a transcendent meaning to their exile.

22. Chazan, *European Jewry*, 252.

23. Ibid., 256.

24. Stow, "Conversion, Apostasy and Apprehensiveness," 932, n.74.

25. Chazan, *European Jewry*, 137.

26. Haverkamp, *Medieval Germany*, 343–44.

27. Ibid., 344. On the civic integration of Jews, see Haverkamp, " 'Concivilitas' von Christen und Juden in Aschkenas im Mittelalter," *Aschkenas* 3 (1996): 103–36.

28. Marcus, "A Jewish-Christian Symbiosis," 486. See now, Peter Schäfer, "Jews and Christians in the High Middle Ages: *The Book of the Pious*," in Christoph Cluse, ed., *The Jews of Europe in the Middle Ages: Tenth to Fifteenth Centuries* (Turnhout, Belgium, 2004), 29–42. "*The Book of the Pious* allows an unprecedented look at the culture of Jewish daily life in the High Middle Ages in the heart of Europe, a culture that in virtually all its aspects was

closely intertwined with that of its Christian surroundings. The Ashkenazi Jews of the Rhineland did not live in a ghetto, but participated in a common cultural space that they shared with their Christian neighbours. Certainly the Jews were in the minority, and their minority culture was dominated by a Christian majority culture. And yet it would be completely false to rigidly separate the two cultures from each other, as many of the examples discussed have shown. The Jewish minority culture and the Christian majority culture were neither hermetically sealed off from each other, nor did the two define themselves exclusively or even primarily in opposition to one another" (41).

29. Marcus, "A Jewish-Christian Symbiosis," 486.

30. Chazan, *European Jewry,* 202–3. And Haverkamp, "Baptised Jews," in Signer and Van Engen, eds., *Jews and Christians,* 255–310.

31. Morrison, *Conversion and Text,* 78, 81, 91, and 96.

32. Robin R. Mundill, *England's Jewish Solution: Experiment and Expulsion, 1262–1290* (Cambridge, 1998), 24–26.

33. Ibid., 33.

34. Ibid., 21–22; 38–39.

35. Ibid., 263: "In 1286, a Jewish wedding took place in Hereford to which Christians had been invited. This event caused Richard Swinfield, bishop of Hereford, to write to the chancellor of Hereford Cathedral ordering him to forbid all Christians from attending the convivialities of the Jews by a proclamation to be made in all the churches in Hereford. Many Christian townsmen clearly ignored the prohibition. After the event Bishop Swinfield, in more caustic mood, claimed that certain numbers of his flock had attended the 'displays of silk and cloth of gold, horsemanship, equestrian processions, stage-playing and sports and minstrelsy' that had accompanied the Jewish wedding feast. Furthermore, he claimed that his congregation had eaten, drunk, played, and jested with the Jews. He warned that all members of the Christian faith who had attended the celebrations should receive absolution within eight days or be excommunicated."

36. Ibid., 34–35.

37. Ibid., 245. Some of those relationships between Jews and Christians, even in the course of lending money, could be positive. See Joseph Shatzmiller, *Shylock Reconsidered: Jews, Moneylending, and Medieval Society* (Berkeley, 1990).

38. Mundill, *England's Jewish Solution,* 246–47.

39. Ibid., 42.

40. Jordan, *French Monarchy,* 33–34.

41. Ibid., 77.

42. Ibid., 152.

43. Ibid. On Béziers and anti-Jewish laws, 150. On Pamiers, 206.

CHAPTER FIVE

VIOLENCE

1. Jews could, of course, be wrong about the nature of threats facing them, as discussed in William Jordan, "Home Again: The Jews in the Kingdom of France, 1315–1322," in F. R. P. Akehurst and Stephanie Van D'Elden, eds., *The Stranger in Medieval Society* (Minneapolis, 1997), 27–43.

2. On violence in medieval society, see Richard W. Kaeuper, ed., *Violence in Medieval Society* (Rochester, NY, 2000); Claude Gauvard, "Fear of Crime in Late Medieval France," in Barbara A. Hanawalt and David Wallace, eds., *Medieval Crime and Social Control* (Minneapolis, 1999), 1–49; William Ian Miller, *Bloodtaking and Peacemaking* (Chicago, 1990); Daniel Lord Smail, "Hatred as a Social Institution in Late-Medieval Society," *Speculum* 76 (2001): 90–126, and Smail, *The Consumption of Justice: Emotions, Publicity, and Legal Culture in Marseille, 1264–1423* (Ithaca, 2003); Barbara H. Rosenwein, ed., *Anger's Past: The Social Uses of an Emotion in the Middle Ages* (Ithaca, 1998); and Mark D. Meyerson, Daniel Thiery, and Oren Falk, eds., *'A Great Effusion of Blood'? Interpreting Medieval Violence* (Toronto, 2004).

3. John Hudson, *The Formation of the English Common Law: Law and Society in England from the Norman Conquest to Magna Carta* (London, 1996), 52. On the search for order amidst violence, see Michael E. Goodich, *Violence and Miracle in the Fourteenth Century: Private Grief and Public Salvation* (Chicago, 1995).

4. See Jody Enders, *The Medieval Theater of Cruelty: Rhetoric, Memory, and Violence* (Ithaca, 1999); Mitchell B. Merback, *The Thief, the Cross, and the Wheel: Pain and the Spectacle of Punishment in Medieval and Renaissance Europe* (Chicago, 1999); Esther Cohen, " 'To Die a Criminal for the Public Good': The Execution Ritual in Late Medieval Paris," in Bernard S. Bachrach and David Nicholas, eds., *Law, Custom, and the Social Fabric in Medieval Europe: Essays in Honor of Bryce Lyon Studies in Medieval Culture* (Kalamazoo, 1990), 285–304.

5. Thomas Bisson, *Tormented Voices: Power, Crisis, and Humanity in Rural Catalonia, 1140–1200* (Cambridge, MA, 1998).

6. John Gilchrist, "The Perception of Jews in the Canon Law in the Period of the First Two Crusades," *Jewish History* 3 (1988): 9–24. For the derogatory Christian iconography of Jews, see Debra Higgs Strickland, *Saracens, Demons, and Jews: Making Monsters in Medieval Art* (Princeton, 2003); Sara Lipton, *Images of Intolerance: The Representation of Jews and Judaism in the* Bible Moraliseé (Berkeley, 1999); and Ruth Mellinkoff, *Outcasts: Signs of Otherness in Northern European Art of the Late Middle Ages,* 2 vols. (Berkeley, 1991).

7. Paul Meyvaert, " 'Rainaldus est malus scriptor Francigenus'—Voicing National Antipathy in the Middle Ages," *Speculum* 66 (1991): 743–64. On the interplay of different identities see, John McLoughlin, "Nations and Loyalties: The Outlook of a Twelfth-Century Schoolman (John of Salisbury, c. 1120–1180)," in David Loades and Katherine Walsh, eds., *Faith and Identity: Christian Political Experience* (Oxford, 1990), 39–47. For an excellent discussion of ethnic hostility—but also the overcoming of such feelings—see Hugh Thomas, *The English and the Normans: Ethnic Hostility, Assimilation, and Identity 1066–c.1220* (Oxford, 2003).

8. G. R. Evans, *Alan of Lille: The Frontiers of Theology in the Late Twelfth Century* (Cambridge, 1983), 124.

9. See John V. Tolan, *Saracens: Islam in the Medieval European Imagination* (New York, 2002); Tolan, ed., *Medieval Christian Perceptions of Islam: A Book of Essays* (New York, 1996); Richard Southern, *Western Views of Islam in the Middle Ages* (Cambridge, MA, 1962); Norman Daniel, *Islam and the West: The Making of an Image* (Oxford, 1993); and David R. Blanks and Michael Frassetto, eds. *Western Views of Islam in Medieval and Early Modern Europe* (New York, 1999).

10. See Paul Freedman, *Images of the Medieval Peasant* (Stanford, 1999), although Freedman is careful to illustrate that there could be positive images of the peasant as well.

11. Abulafia, *Christians and Jews,* 123. Abulafia herself has shown how varied Christian responses were to the Jews. See Abulafia, "The Intellectual and Spiritual Quest for Christ and Central Medieval Persecution of Jews," in her collection, *Religious Violence Between Christians and Jews,* 61–85. On whether invective against the Jews later in the Middle Ages represented rhetorical violence of a different order compared with other polemical disputes, see Rubenstein, *Guibert of Nogent,* 116–24.

12. Simonsohn, *Apostolic See,* 7:17–21, on Innocent III, and on *sicut judeis,* 44–45.

13. Jeremy Cohen, *Friars and the Jews: The Evolution of Medieval Anti-Judaism* (Ithaca, 1982).

14. See, for example, Penn R. Szittya, *The Antifraternal Tradition in Medieval Literature* (Princeton, 1986).

15. Stefan Rohrbacher, "The Charge of Deicide. An Anti-Jewish Motif in Medieval Christian Art," *Journal of Medieval History* 17 (1991): 313. See, too, Jordan, "The Erosion of the Stereotype of the Last Tormentor of Christ," *Jewish Quarterly Review* 81 (1990): 13–44.

16. Chazan, *European Jewry*, 203.

17. Ibid., 201–2. For an excellent discussion, see Marcus, "The Dynamics of Jewish Renaissance," 27–46.

18. Anthony Bayle, "Fictions of Judaism before 1290," in Patricia Skinner, ed., *Jews in Medieval Britain: Historical, Literary, and Archaeological Perspectives* (Rochester, NY, 2003), 133.

19. Langmuir, "The Knight's Tale of Young Hugh of Lincoln," in his *Toward a Definition of Antisemitism*, 241.

20. Ibid., 248.

21. Langmuir, "Thomas of Monmouth: Detector of Ritual Murder," *in Toward a Definition of Antisemitism*, 209–236, especially 221. See now, John M. McCulloh, "Jewish Ritual Murder: William of Norwich, Thomas of Monmouth, and the Early Dissemination of the Myth," *Speculum* 72 (1997): 698–740. McCulloh offers a revision of Langmuir's analysis, arguing that some basic elements of the ritual murder accusation must have existed before Monmouth's literary creation. In any event, the absence of enthusiasm for William's cult in Norwich (even if word of the cult had spread to a few locations in Germany) suggests that the polemical violence against the Jews did not promote physical violence against them.

22. Bayle, "Fictions of Judaism before 1290," 131.

23. Langmuir, "The Knight's Tale, 240.

24. Ibid., 240.

25. Ibid., 241.

26. Ibid., 241. See also Langmuir, "Historiographic Crucifixion," in *Toward a Definition of Antisemitism*, 284, for a discussion of the growth of the crucifixion aspect of the murder accusation.

27. See Solomon Grayzel, *The Church and the Jews in the XIIIth Century*, vol. 2, *1254–1314*, Kenneth R. Stow, ed., (Detroit, 1989), 198, n.2, citing the statistics compiled by Peter Browe.

28. Robert Stacey, "Crusades, Martyrdoms, and the Jews of Norman England, 1096–1190," in Alfred Haverkamp, ed., *Juden und Christen im Zeitalter der Kreuzüge,* 234.

29. Stacey, "Crusades," 242–43 and 246–51.

30. Barrie Dobson, "The Medieval York Jewry Reconsidered," in Skinner, ed., Jews in *Medieval Britain: Historical, Literary, and Archeological Perspectives.*

31. Dobson, "York Jewry Reconsidered," 148.

32. Jordan, *French Monarchy,* 115.

33. Ibid., 122.

34. Ibid., 124.

35. Ibid., 147–48.

36. Nirenberg, *Communities of Violence,* 71.

37. Jordan, *French Monarchy,* 244.

38. Ibid., 245. Jordan here describes the course of anti-Judaism in Chinon as building progressively to the massacre. "Bitterness over owing money to Jews (and, by transference, toward Jews themselves) became the common denominator of Christian society in Chinon." To describe anti-Judaism as the common denominator of the town is perhaps too extreme a conclusion based only on the vagaries of medieval evidence.

39. Ibid., 246.

40. Much of the following discussion is based on Friederich Lotter's account in Norman Roth, *Medieval Jewish Civilization: An Encyclopedia,* 293–304.

41. Langmuir, "Ritual Cannibalism," in *Toward a Definition of Antisemitism,"* 263–81.

42. Ibid., 265.

43. Rubin, "Imagining the Jew: The Late Medieval Eucharistic Discourse," in R. Po-Chia Hsia and Hartmut Lehmann, eds., *In and Out of the Ghetto: Jewish-Gentile Relations in Late Medieval and Early Modern Germany* (Washington, D.C., 1995), 189–90.

44. Rubin, *Gentile Tales,* 55.

45. Ibid., 55.

46. Rubin, "Imagining the Jew," 193–94.

47. Ibid., 197 ("In 1337, the townspeople would withstand the attack of the Armleder mob.")

48. Ibid., 197.

49. Ibid., 198.

50. Ibid., 198.

51. Ibid., 202.

52. Rubin, *Gentile Tales*, 56.

53. Ibid., 54.

54. Haverkamp, "The Jewish Quarters in German Towns during the Late Middle Ages," in Hsia and Lehmann, eds., *In and Out of the Ghetto*, 16.

55. Marcus, "A Jewish-Christian Symbiosis," 459–60.

56. Bell, *Sacred Communities*, 116. See Haverkamp, "Die Judenverfolgungen zur Zeit des Schwarzen Todes in Gesellschaftsgefüge deutscher Staädte," in Haverkamp, ed., *Zur Geschichte der Juden im Deutschland des Späten Mittelalters und der Frühen Neuzeit* (Stuttgart, 1981), 27–93.

57. Baron, *Social and Religious History*, 9:207. Jonathan Israel, *European Jewry in the Age of Mercantilism 1550–1750*, 3rd. ed. (London, 1998), 5: "Vibrant communities formed again in Augsburg, Nuremberg, Ulm, Mainz, Worms and many other cities where the Jews had temporarily been all but destroyed."

58. Baron, *Social and Religious History*, 10:10.

59. Ibid., 10:18.

60. Ibid., 10:38.

61. Ibid., 9:231.

62. Ibid., 9:212.

63. Ibid., 9:218.

64. Ibid., 9:227.

65. Hsia, *The Myth of Ritual Murder: Jews and Magic in Reformation Germany* (New Haven, 1988), 88.

66. Haverkamp, "Jewish Quarters," 23.

67. Nirenberg, *Communities of Violence*, 74.

68. Ibid., 91.

69. Ibid., 223: "What is most conspicuous about Holy Week violence is its limits. In town after town, year after year, crowds of children hurled stones and insults at Jews and the homes of Jews without inciting broader riot."

70. Ibid., 237.

71. Gampel, "Jews, Christians, and Muslims in Medieval Iberia: *Convivencia* through the Eyes of Sephardic Jews," in Vivian B. Mann, Thomas F. Glick, and Jerrilynn D. Dodds, eds., Convivencia: *Jews, Muslims, and Christians in Medieval Spain* (New York, 1992), 26.

72. Ibid., 28.

73. Philippe Wolff, "The 1391 Pogrom in Spain: Social Crisis or Not?" *Past and Present* 50 (1971): 4–18. See Michel Mollat and Philippe Wolff, *The Popular Revolutions of the Late Middle Ages* (London, 1973).

74. Wolff, "1391," 8.

75. Ibid., 15.

76. Gampel, "*Convivencia,*" 29.

77. Gampel, "Does Medieval Navarrese Jewry Salvage Our Notion of *Convivencia?*" In Bernard Dov Cooperman, ed., *In Iberia and Beyond: Hispanic Jews between Cultures* (Newark, DE, 1998), 98.

78. Mark Meyerson, *A Jewish Renaissance in Fifteenth-Century Spain* (Princeton, 2004).

79. Gampel, "*Convivencia,*" 30.

80. See Gretchen D. Starr-LeBeau, *In the Shadow of the Virgin: Inquisitors, Friars, and Conversos in Guadalupe, Spain* (Princeton, 2003), 89.

CHAPTER SIX
EXPULSION AND CONTINUITY

1. Mundill, *England's Jewish Solution*, 281.

2. Robert Stacey, "Parliamentary Negotiation and the Expulsion of the Jews from England," in Michael Prestwich, R. H. Britnell, and Robin Frame, eds., *Thirteenth Century England VI* (Woodbridge, 1997), 101.

3. Jordan, *French Monarchy*, 253.

4. Ibid., 242.

5. Ibid., 209.

6. Ibid., 248–50.

7. The touchstone article is Joseph Strayer, "France: The Holy Land, the Chosen People, and the Most Christian King," in Strayer, *Medieval Statecraft and the Perspectives of History* (Princeton, 1971), 300–15. See also Jordan's discussion of the precocious nature of Capetian state-building in *French Monarchy*, 128–29, and his reservations on 252–53.

8. Jordan, *French Monarchy*, 246.

9. See the provocative challenge to the scholarly enthusiasm for the medieval state in Rees Davies, "The Medieval State: The Tyranny of a Concept?" *Journal of Historical Sociology* 16 (2003): 280–300.

10. Edward Peters, "Jewish History and Gentile Memory: The Expulsion of 1492," *Jewish History* 9 (1995): 9–33.

11. See the introduction to Thomas A. Brady, Jr., Heiko Oberman, James D. Tracy, eds., *Handbook of European History, 1400–1600: Late Middle Ages, Renaissance and Reformation,* 2 vols. (Leiden, 1994), 1:xvi.

12. Brady et al., eds., *Handbook,* 1:xvi–xvii.

13. Ibid., 1:xvii–xx.

14. Israel, *European Jewry,* 29–43.

15. Scribner, "Preconditions of Tolerance and Intolerance in Sixteenth-Century Germany," in Ole Peter Grell and Bob Scribner, eds., *Tolerance and Intolerance in the European Reformation* (Cambridge, 1996), 38.

16. Ibid., 44–45.

17. For a nice discussion of debates about the integration of Jews in Italy and the interaction of Jewish internal culture with that of Italian society, see David Ruderman, "Cecil Roth, Historian of Italian Jewry: A Reassessment," in David N. Myers and David B. Ruderman, eds., *The Jewish Past Revisited: Reflections on Modern Jewish Historians* (New Haven, 1998), 128–43, at 139: "No doubt to reduce a cultural profile of the Jewish minority to mere 'servile imitation' of the majority culture is distorting and flattens Jewish culture to a mere set of responses to the Other, either positive or negative. But the opposite extreme, that of viewing a minority civilization from a purely internalist perspective, as creating its own culture in its own terms, is also distorting. It is rather the negotiation of the inside with the outside that correctly constitutes the proper focus of the historian's gaze."

18. Anna Foa, *The Jews of Europe after the Black Death* (Berkeley, 2000) 117–21.

19. Michele Luzzati, "Northern and Central Italy: Assessment of Research and Further Prospects," in Cluse, ed., *Jews of Europe in the Middle Ages,* 194.

20. Foa, *Jews,* 124.

21. Stefanie B. Siegmund, *The Medici State and the Ghetto of Florence: The Construction of an Early Modern Jewish Community* (Stanford, 2006).

22. Robert Bonfil, *Jewish Life in Renaissance Italy* (Berkeley, 1994), 71–73.

23. Foa, *Jews,* 150. See also Kenneth Stow, *Theater of Acculturation: The Roman Ghetto in the 16th Century* (Seattle, 2001).

24. Foa, *Jews,* 149.

25. Ibid., 150.

26. Mark R. Cohen, ed. and trans., *The Autobiography of a Seventeenth-Century Venetian Rabbi: Leon of Modena's Life of Judah* (Princeton, 1988).

27. Michael Toch, "Population History and Village Jews," in Hsia and Lehmann, eds., *Ghetto,* 82.

28. Bell, *Sacred Communities*, 208.

29. Ibid., 86.

30. Haverkamp, "Jewish Quarters," 22.

31. Bell, *Sacred Communities*, 88.

32. Ibid., 108.

33. Hsia, *Ritual Murder*, 81–82.

34. Ibid., 94.

35. Ibid., 205. Christians could sometimes come to more benign visions of the Jews, even in local productions of passion plays. See John D. Martin, "Dramatized Disputations: Late Medieval German Dramatizations of Jewish-Christian Religious Disputations, Church Policy, and Local Social Climates," *Medieval Encounters* 8 (2002): 209–27.

36. Hsia, *Ritual Murder*, 147.

37. See, too, Yacov Guggenheim, "Meeting on the Road: Encounters between German Jews and Christians on the Margins of Society," in Hsia and Lehmann, eds., *Ghetto*, 125: "As a result, by the second half of the fourteenth century only approximately seven thousand Jewish families, that is, between 25,000 and 30,000 Jewish individuals, were left in the realm of the Holy Roman Empire. The demographic figures remained stable until the beginning of the sixteenth century, when for economic reasons, the natural surplus of the Jewish population emigrated to the south and to the east."

38. Christoph Daxelmüller, "Organizational Forms of Jewish Popular Culture since the Middle Ages," in Hsia and Lehmann, eds., *Ghetto*, 32.

39. Bell, *Sacred Communities*, 4.

40. Rotraud Ries, "German Territorial Princes and the Jews," in Hsia and Lehmann, eds., *Ghetto*, 267.

41. J. Friedrich Battenberg, "Jews in Ecclesiastical Territories of the Holy Roman Empire," in Hsia and Lehmann, eds., *Ghetto*, 252: "Since the late sixteenth century, however, ecclesiastical princes were much more prepared to tolerate Jews. . . . It is striking that after the 1570s the Jews were readmitted to many ecclesiastical states but not to a single secular one." Ries, "German Territorial Princes and the Jews," 267.

42. Lotter, in Roth, ed., *Medieval Jewish Civilization: An Encyclopedia*, 303.

43. Haverkamp, "Jewish Quarters," 17.

44. Bell, *Sacred Communities*, 91.

45. Israel, "Germany and Its Jews (1300–1800)," in Hsia and Lehmann, eds., *Ghetto*, 297.

46. Thomas Brady, "Germans with a Difference? The Jews of the Holy Roman Empire during the Early Modern Era—A Comment," in Hsia and Lehmann, eds., *Ghetto*, 290. On Anabaptists, see Hartmut Lehmann, "The Jewish Minority and the Christian Majority in Early Modern Central Europe," in Hsia and Lehmann, eds., *Ghetto*, 306.

47. Foa, *Jews*, 159–60.

CONCLUSION

1. Glick, "*Convivencia*: An Introductory Note," in Mann et al., eds., *Convivencia*, 1. *Convivencia* as a term of analysis is now being rethought. See Jonathan Ray, "Beyond Tolerance and Persecution: Reassessing Our Approach to Medieval *Convivencia*," *Jewish Social Studies* 11 (winter 2005): 1–18; and Alex Novikoff, "Between Tolerance and Intolerance in Medieval Spain: An Historiographical Enigma," *Medieval Encounters* 11 (2005): 7–36.

2. Glick, "*Convivencia*," 4.

3. Ibid., 4.

4. Benjamin Gampel, "Jews, Christians, and Muslims," in Mann et al., eds., *Convivencia*, 11.

5. Ibid., 21.

6. Diarmaid MacCulloch, *The Reformation: A History* (London, 2003), 185–86.

BIBLIOGRAPHY

Abulafia, Anna Sapir. *Christians and Jews in the Twelfth-Century Renaissance.* London, 1995.

———, ed. *Religious Violence between Christians and Jews: Medieval Roots, Modern Perspectives.* New York, 2002.

Albert, Bat-Sheva. "Adversus Iudaeos in the Carolingian Empire." In Ora Limor and Guy G. Stroumsa, eds., *Contra Iudaeos: Ancient and Medieval Polemics between Christians and Jews.* Tübingen, 1996. 119–42.

———. "Isidore of Seville: His Attitude towards Judaism and His Impact on Early Canon Law." *Jewish Quarterly Review* 80 (1990): 207–20.

Amory, Patrick. *People and Identity in Ostrogothic Italy, 489–554.* Cambridge, 1997. 13–42.

The Annals of St-Bertin. Translated by Janet L. Nelson. Manchester, 1991.

Bachrach, Bernard S. *Early Medieval Jewish Policy in Western Europe.* Minneapolis, 1977.

Baron, Salo. "Ghetto and Emancipation." *Menorah Journal* 14 (1928): 515–26.

———. *A Social and Religious History of the Jews.* 2nd ed. 18 vols. New York, 1952–.

Bartlett, Robert. *The Making of Europe: Conquest, Colonization, and Cultural Change, 950–1350.* Princeton, 1993.

Battenberg, J. Friederich. "Jews in Ecclesiastical Territories of the Holy Roman Empire." In R. Po-Chia Hsia and Hartmut Lehmann, eds., *In and Out of the Ghetto: Jewish-Gentile Relations in Late Medieval and Early Modern Germany.* Washington, D.C., 1995. 247–76.

Bell, Dean Phillip. *Sacred Communities: Jewish and Christian Identities in Fifteenth-Century Germany.* Leiden, 2001.

Benin, Stephen D. "Jews, Muslims, and Christians in Byzantine Italy." In Benjamin H. Hary, John L. Hayes, Fred Astren, eds., *Judaism and Islam: Boundaries, Communication, and Interaction: Essays in Honor of William M. Brinner.* Leiden, 2000. 27–35.

Bernstein, Michael André. *Foregone Conclusions: Against Apocalyptic History.* Berkeley, 1994.

Biale, David, ed. *Cultures of the Jews: A New History.* New York, 2002.

Bisson, Thomas. *Tormented Voices: Power, Crisis, and Humanity in Rural Catalonia, 1140–1200.* Cambridge, MA, 1998.

Blanks, David R., and Michael Frassetto, eds. *Western Views of Islam in Medieval and Early Modern Europe.* New York, 1999.

Blumenkranz, Bernard. *Juifs et Chrétiens dans le monde occidental 430–1096.* Paris, 1960.

Bonfil, Robert. "The Cultural and Religious Traditions of French Jewry in the Ninth Century, as Reflected in the Writings of Agobard of Lyons." In Joseph Dan and Joseph Hacker, eds., *Studies in Jewish Mysticism, Philosophy, and Ethical Literature Presented to Isaiah Tishby on his Seventy-fifth Birthday.* Jerusalem, 1986. 327–49. [Hebrew].

———. *Jewish Life in Renaissance Italy.* Berkeley, 1994.

Bradbury, Scott, ed. and trans. *Severus of Minorca, Letter on the Conversion of the Jews.* Oxford, 1996.

Bradshaw, Paul F., and Lawrence A. Hoffman, eds. *Passover and Easter: Origin and History to Modern Times.* Notre Dame, IN, 1999.

Brady, Jr., Thomas A. "Germans with a Difference? The Jews of the Holy Roman Empire during the Early Modern Era—A Comment." In R. Po-Chia Hsia and Hartmut Lehmann, eds., *In and Out of the Ghetto: Jewish-Gentile Relations in Late Medieval and Early Modern Germany.* Washington, D.C., 1995. 289–94.

Brady, Jr., Thomas A., Heiko Oberman, and James D. Tracy, eds. *Handbook of European History, 1400–1600: Late Middle Ages, Renaissance and Reformation. Structures and Assertions.* 2 vols. Leiden, 1994.

Brennan, Brian. "The Conversion of the Jews of Clermont in AD 576." *Journal of Theological Studies* 36 (1985): 321–37.

Brown, Peter R. L. *The Cult of the Saints: Its Rise and Function in Latin Christianity.* Chicago, 1981.

———. *The Rise of Western Christendom: Triumph and Diversity, A.D. 200–1000.* Cambridge, MA, 1996.

Bynum, Caroline Walker. *Jesus as Mother: Studies in the Spirituality of the High Middle Ages.* Berkeley, 1982.

Carlebach, Elishevah. "The Sabbatian Posture of German Jewry." *Jerusalem Studies in Jewish Thought* 16–17 (2001): 1–29.

Cavadini, John C., ed. *Gregory the Great: A Symposium.* Notre Dame, IN, 1995.

Chazan, Robert. "1007–1012: Initial Crisis for Northern-European Jewry." *Proceedings of the American Academy for Jewish Research* 37–39 (1970–1971): 101–17.

———. "The Anti-Jewish Violence of 1096: Perpetrators and Dynamics." In Anna Sapir Abulafia, ed., *Religious Violence between Christians and Jews: Medieval Roots, Modern Perspectives.* New York, 2002. 21–43.

———. *Church, State, and Jew in the Middle Ages.* New York, 1980.

———. *European Jewry and the First Crusade.* Berkeley, 1987.

Chazelle, Celia. *The Crucified God in the Carolingian Era: Theology and Art of Christ's Passion.* Cambridge, 2001.

The Chronicle of Ahimaaz. Translated and edited by Marcus Salzman. New York, 1966.

Chronicle of Fredegar. Translated by J. M. Wallace-Hadrill. London, 1960.

Cluse, Christoph. *Studien zur Geschichte der Juden in dem mitteralterichen Niederlanden.* Hanover, 2000.

———, ed. *The Jews of Europe in the Middle Ages: Tenth to Fifteenth Centuries.* Turnhout, Belgium, 2004.

Cohen, Esther. " 'To Die a Criminal for the Public Good': The Execution Ritual in Late Medieval Paris." In Bernard S. Bachrach and David Nicholas, eds., *Law, Custom, and the Social Fabric in Medieval Europe: Essays in Honor of Bryce Lyon.* Kalamazoo, MI, 1990. 285–304.

Cohen, Jeremy. "Between Martyrdom and Apostasy: Doubt and Self-Definition in Twelfth-Century Ashkenaz." *Journal of Medieval and Early Modern Studies* 29 (1999): 431–71.

———. *Friars and the Jews: the Evolution of Medieval Anti-Judaism.* Ithaca, NY, 1982.

———. "The Jews as the Killers of Christ in the Latin Tradition, from St. Augustine to the Friars." *Traditio* 39 (1983): 1–27.

———. *Living Letters of the Law: Ideas of the Jew in Medieval Christianity.* Berkeley, 1999.

Cohen, Mark, R. *Under Crescent and Cross: The Jews in the Middle Ages.* Princeton, 1994.

———, ed. and trans. *The Autobiography of a Seventeenth-Century Venetian Rabbi: Leon of Modena's* Life of Judah. Princeton, 1988.

Constable, Giles. *The Reformation of the Twelfth Century.* Cambridge, 1996.

Dahan, Gilbert. *Les Intellectuals Chrétiens et les Juifs au Moyen Age.* Paris, 1990.

Daniel, Norman. *Islam and the West: The Making of an Image.* Oxford, 1993.

Davies, Rees. "The Medieval State: The Tyranny of a Concept?" *Journal of Historical Sociology* 16 (2003): 280–300.

De Jonge, Mayke. "The Empire as *Ecclesia*: Hrabanus Maurus and Biblical *Historia* for Rulers." In Yitzhak Hen and Matthew Innes, eds., *The Uses of the Past in the Early Middle Ages.* Cambridge, 2000. 191–227.

———. "Exegesis for an Empress." In Esther Cohen and Mayke de Jonge, eds., *Medieval Transformations: Texts, Power, and Gifts in Context.* Leiden, 2001. 69–101.

———. "Old Law and New-Found Power: Hrabanus Maurus and the Old Testament." In J. W. Drijvers and A. A. MacDonald, eds., *Centres of Learning: Learning and Location in Pre-Modern Europe and the Near East.* Leiden, 1995. 161–76.

De Nie, Giselle. *Views from a Many-windowed Tower: Studies in the Imagination in the Works of Gregory of Tours.* Amsterdam, 1987.

Diaz, Pablo, and Ma R. Valverde. "The Theoretical Strength and Practical Weakness of the Visigothic Monarchy of Toledo." In Frans Theuws and Janet L. Nelson, eds., *Rituals of Power: From Late Antiquity to the Early Middle Ages.* Leiden, 2000. 59–94.

DuQuesnay Adams, Jeremy. "Ideology and the Requirements of 'Citizenship' in Visigothic Spain: The Case of the *Judaei.*" *Societas* 2 (1972): 317–32.

Eidelberg, Shlomo. *The Jews and the Crusaders: The Hebrew Chronicles of the First and Second Crusades.* Madison, WI, 1977.

Elukin, Jonathan. "The Discovery of the Self: Jews and Conversion in the Twelfth Century." In John Van Engen and Michael Signer, eds., *Jews and Christians in Twelfth-Century Europe.* Notre Dame, IN, 2001. 63–77.

———. "Jacques Basnage and the *History of the Jews*: Polemic and Allegory in the Republic of Letters." *Journal of the History of Ideas* 53 (1992): 603–31.

———. "From Jew to Christian? Conversion and Immutability in Medieval Europe." In James Muldoon, ed., *Varieties of Religious Conversion in the Middle Ages.* Miami, 1997. 171–90.

———. "Judaism: From Heresy to Pharisee in Medieval Christian Literature." *Traditio* 57 (2002): 49–66.

———. "A New Essenism: Heinrich Graetz and Mysticism." *Journal of the History of Ideas* 59 (1998): 135–48.

Enders, Jody. *The Medieval Theater of Cruelty: Rhetoric, Memory, and Violence.* Ithaca, NY, 1999.

Epstein, Marc Michael. *Dreams of Subversion in Medieval Jewish Art and Literature.* University Park, PA, 1997.

Evans, G. R. *Alan of Lille: The Frontiers of Theology in the Late Twelfth Century.* Cambridge, 1983.

Feiner, Shmuel. Haskalah *and History: The Emergence of a Modern Jewish Historical Consciousness.* Oxford, 2002.

Feldman, Louis. *Jew and Gentile in the Ancient World: Attitudes and Interactions from Alexander to Justinian.* Princeton, 1993.

Foa, Anna. *The Jews of Europe after the Black Death.* Berkeley, 2000.

Fouracre, Paul, ed. *The New Cambridge Medieval History.* Vol. 1. *c. 500–c.700.* Cambridge, 2000.

Fredriksen, Paula. "*Secundem Carnem:* History and Israel in the Theology of St. Augustine." In William E. Klingshirn and Mark Vessey, eds., *The Limits of Ancient Christianity: Essays on Late Antique Thought and Culture in Honor of R. A. Markus.* Ann Arbor, 1999. 26–42.

Freedman, Paul. *Images of the Medieval Peasant.* Stanford, 1999.

———. "The Medieval 'Other.' The Middle Ages as Other." In Timothy S. Jones and David A. Sprunger, eds., *Marvels, Monsters, and Miracles: Studies in the Medieval and Early Modern Imagination.* Kalamazoo, 2002. 1–27.

Freedman, Paul, and Gabrielle Spiegel. "Medievalisms Old and New: The Rediscovery of Alterity in North American Medieval Studies." *American Historical Review* 103 (1998): 677–704.

Gampel, Benjamin R. "Does Medieval Navarrese Jewry Salvage Our Notion of *Convivencia*?" In Bernard Dov Cooperman, ed., *In Iberia and Beyond: Hispanic Jews between Cultures.* Newark, NJ, 1998. 97–122.

———. "Jews, Christians, and Muslims in Medieval Iberia: *Convivencia* through the Eyes of Sephardic Jews." In Vivian B. Mann, Thomas F. Glick, and Jerrilynn D. Dodds, eds., Convivencia: *Jews, Muslims, and Christians in Medieval Spain.* New York, 1992. 11–38.

———, ed. *Crisis and Creativity in the Sephardic World: 1391–1648.* New York, 1997.

Gauvard, Claude. "Fear of Crime in Late Medieval France." In Barbara A. Hanawalt and David Wallace, eds., *Medieval Crime and Social Control.* Minneapolis, 1999. 1–49.

Gervers, Michael, and James M. Powell, eds. *Tolerance and Intolerance: Social Conflict in the Age of the Crusades.* Syracuse, 2001.

Gilchrist, John. "The Perception of Jews in the Canon Law in the Period of the First Two Crusades." *Jewish History* 3 (1988): 9–24.

Gilman, Sander L. *Jewish Self-Hatred*. Baltimore, 1986.

Ginzburg, Carlo. "The Conversion of the Minorcan Jews (417–18): An Experiment in History of Historiography." In Scott L. Waugh and Peter D. Diehl, eds., *Christendom and Its Discontents: Exclusion, Persecution, and Rebellion, 1000–1500*. Cambridge, 1996. 207–19.

Goffart, Walter. *The Narrators of Barbarian History: AD 550–800, Jordanes, Gregory of Tours, Bede, and Paul the Deacon*. Princeton, 1988.

Golb, Norman. *The Jews in Medieval Normandy: A Social and Intellectual History*. Cambridge, 1998.

Goldhill, Simon. *The Temple of Jerusalem*. Cambridge, MA, 2005.

Gonzalez-Salinero, R. "Catholic Anti-Judaism in Visigothic Spain." In Albert Ferreiro, ed., *The Visigoths: Studies in Culture and Society*. Leiden, 1999. 123–51.

Goodich, Michael. *Violence and Miracle in the Fourteenth Century: Private Grief and Public Salvation*. Chicago, 1995.

Grayzel, Solomon. *The Church and the Jews in the XIIIth Century*. Vol. 2. *1254–1314*. Edited by Kenneth R. Stow. Detroit, 1989.

Green, Arthur. "*Shekhinah*, the Virgin Mary, and the *Song of Songs*: Reflections on a Kabbalistic Symbol in Its Historical Context." *Association for Jewish Studies Review* 26 (2002): 1–52.

Gregory the Great. *Saint Gregory the Great Dialogues*. Translated by Odo John Zimmerman, O.S.B., in *Fathers of the Church*. Vol. 39. New York, 1959.

Gregory of Tours. *Glory of the Martyrs*. Translated by Raymond Van Dam. Liverpool, 1988.

———. *The History of the Franks*. Translated by Lewis Thorpe. Harmondsworth, 1974.

Gruen, Erich. *Diaspora: Jews Amidst Greeks and Romans*. Cambridge, MA, 2002.

Guggenheim, Yacov. "Meeting on the Road: Encounters between German Jews and Christians on the Margins of Society." In R. Po-Chia Hsia and Hartmut Lehmann, eds., *In and Out of the Ghetto: Jewish-Gentile Relations in Late Medieval and Early Modern Germany*. Washington, D.C., 1995. 125–36.

Guibert of Nogent. *A Monk's Confession: The Memoirs of Guibert of Nogent*. Translated by Paul J. Archambault. University Park, PA, 1996.

Haverkamp, Alfred. " 'Concivilitas' von Christen und Juden in Aschkenas im Mittelalter." *Aschkenas* 3 (1996): 103–36.

————. "The Jewish Quarters in German Towns during the Late Middle Ages." In R. Po-Chia Hsia and Hartmut Lehmann, eds., *In and Out of the Ghetto: Jewish-Gentile Relations in Late Medieval and Early Modern Germany.* Washington, DC, 1995. 13–28.

————. "Die Judenverfolgungen zur Zeit des Schwarzen Todes in Gesellschaftsgefüge deutscher Staädte." In Alfred Haverkamp, ed., *Zur Geschichte der Juden im Deutschland des Späten Mittelalters und der Frühen Neuzeit.* Stuttgart, 1981. 27–93.

————. *Medieval Germany 1056–1273.* Oxford, 1988.

————, ed. *Juden und Christen zur Zeit der Kreuzzüge.* Sigmaringen, 1999.

Heather, Peter. "The Creation of the Visigoths." In Peter Heather, ed., *The Visigoths From the Migration Period to the Seventh Century: An Ethnographic Perspective.* Woodbridge, Suffolk, UK, 1999, 43–92.

Heil, Johannes. " 'Deep Enmity' and/or 'Close Ties?' Jews and Christians before 1096: Sources, Hermeneutics, and Writing History in 1996." *Jewish Studies Quarterly* 9 (2002): 259–306.

————. "Labourers in the Lord's Quarry: Carolingian Exegetes, Patristic Authority, and Theological Innovation: A Case Study in the Representation of Jews in Commentaries on Paul." In Celia Chazelle and Burton Van Name Edwards, eds., *The Study of the Bible in the Carolingian Era.* Turnhout, Belgium, 2003. 75–96.

Heinzelmann, Martin. *Gregory of Tours: History and Society in the Sixth Century.* Cambridge, 2001.

Hen, Yitzhak. *Culture and Religion in Merovingian Gaul, A.D. 481–751.* Leiden, 1995.

————. "Paganism and Superstitions in the Time of Gregory of Tours: Une Question mal posée!" In Kathleen Mitchell and Ian N. Wood, eds., *The World of Gregory of Tours.* Leiden, 2002, 229–40.

Hen, Yitzhak, and Matthew Innes. *The Uses of the Past in the Early Middle Ages.* Cambridge, 2000.

Herrin, Judith. *The Formation of Christendom.* Princeton, 1987.

Hood, John Y. B. *Aquinas and the Jews.* Philadelphia, 1995.

Howard-Johnston, James, and Paul Antony Hayward, eds. *The Cult of Saints in Late Antiquity and the Middle Ages: Essays on the Contribution of Peter Brown.* Oxford, 1999.

Hsia, R. Po-Chia. *The Myth of Ritual Murder: Jews and Magic in Reformation Germany.* New Haven, 1988.

Hsia, R. Po-Chia, and Hartmut Lehmann, eds. *In and Out of the Ghetto: Jewish-Gentile Relations in Late Medieval and Early Modern Germany.* Washington, D.C., 1995.

Hudson, John. *The Formation of the English Common Law: Law and Society in England from the Norman Conquest to Magna Carta.* London, 1996.

Hunt, E. D. "St. Stephen in Minorca: An Episode in Jewish-Christian Relations in the Early 5th Century AD." *Journal of Theological Studies* n.s. 33 (1982): 106–23.

Israel, Jonathan. *European Jewry in the Age of Mercantilism, 1550–1750.* 3rd ed. London, 1998.

James, Edward, ed. *Visigothic Spain: New Approaches.* Oxford, 1980.

Jeep, John, ed. *Medieval Germany: An Encyclopedia.* New York, 2001.

Jones, W. R. "The Image of the Barbarian in Medieval Europe." *Comparative Studies in Society & History* 13 (1971): 376–407.

Jordan, William Chester. "The Erosion of the Stereotype of the Last Tormentor of Christ." *Jewish Quarterly Review* 81 (1990): 13–44.

———. *The French Monarchy and the Jews: From Philip Augustus to the Last Capetians.* Philadelphia, 1989.

———. "Home Again: The Jews in the Kingdom of France, 1315–1322." In F. R. P. Akehurst and Stephanie Van D'Elden, eds., *The Stranger in Medieval Society.* Vol. 12. *Medieval Cultures.* Minneapolis, 1997. 27–43.

Kaeuper, Richard, ed. *Violence in Medieval Society.* Rochester, NY, 2000.

Katz, Solomon. *The Jews in the Visigothic and Frankish Kingdoms of Spain and Gaul.* Cambridge, MA, 1937.

Keely, Avril. "Arians and Jews in the *Histories* of Gregory of Tours." *Journal of Medieval History* 23 (1997): 103–15.

Langmuir, Gavin. "The Tortures of the Body of Christ." In Scott Waugh and Peter Diehl, eds., *Christendom and Its Discontents: Exclusion, Persecution, and Rebellion, 1000–1500.* Cambridge, 1996. 287–309.

———. *Toward a Definition of Antisemitism.* Berkeley, 1990.

Lerner, Robert E. *The Feast of Saint Abraham: Medieval Millenarians and the Jews.* Philadelphia, 2001.

Lieu, Judith, John North, and Tessa Rajak, eds. *The Jews among Pagans and Christians in the Roman Empire.* London, 1992. 174–94.

Linder, Amnon, ed. and trans. *Jews in the Legal Sources of the Early Middle Ages.* Detroit, 1997.

Lipton, Sara. *Images of Intolerance: The Representation of Jews and Judaism in the Bible moralisée.* Berkeley, 1999.

Little, Lester, and Barbara H. Rosenwein, eds. *Debating the Middle Ages: Issues and Readings*. Oxford, 1998.

Lives of the Visigothic Fathers. Translated and edited by A. T. Fear. Liverpool, 1997.

Maccoby, Hyam. *Judaism on Trial: Jewish-Christian Disputations in the Middle Ages*. Rutherford, NJ, 1982.

MacCulloch, Diarmaid. *The Reformation: A History*. New York, 2003.

MacMullen, Ramsay. *Christianity and Paganism in the Fourth to the Eighth Centuries*. New Haven, 1997.

Mann, Vivian, Thomas F. Glick, and Jerrilynn D. Dodds, eds. Convivencia: *Jews, Muslims, and Christians in Medieval Spain*. New York, 1992.

Marcus, Ivan. "The Dynamics of Jewish Renaissance and Renewal in the Twelfth Century." In Michael A. Signer and John Van Engen, eds., *Jews and Christians in Twelfth-Century Europe*. Notre Dame, IN, 2001. 27–45.

———. "A Jewish-Christian Symbiosis: The Culture of Early Ashkenaz." In David Biale, ed., *Cultures of the Jews: A New History*. New York, 2002. 449–515.

———. *Rituals of Childhood: Jewish Acculturation in Medieval Europe*. New Haven, 1996.

Markus, Robert A. *Gregory the Great and His World*. Cambridge, 1997.

Martin, John D. "Dramatized Disputations: Late Medieval German Dramatizations of Jewish-Christian Religious Disputations, Church Policy, and Local Social Climates." *Medieval Encounters* 8 (2002): 209–27.

McCormick, Michael. *Origins of the European Economy: Communications and Commerce, A.D. 300–900*. Cambridge, 2001.

McCulloh, John M. "Jewish Ritual Murder: William of Norwich, Thomas of Monmouth, and the Early Dissemination of the Myth." *Speculum* 72 (1997): 698–740.

McKitterick, Rosamond, ed. *The Early Middle Ages*. Oxford, 2001.

———. *The New Cambridge Medieval History*. Vol. 2. *c. 700–c.900*. Cambridge, 1995.

McLoughlin, John. "Nations and Loyalties: The Outlook of a Twelfth-Century Schoolman (John of Salisbury, c. 1120–1180)." In David Loades and Katherine Walsh, eds., *Faith and Identity: Christian Political Experience*. Oxford, 1990. 39–47.

Mellinkoff, Ruth. *Outcasts: Signs of Otherness in Northern European Art of the Late Middle Ages*. 2 vols. Berkeley, 1991.

Merback, Mitchell B. *The Thief, the Cross, and the Wheel: Pain and the Spectacle of Punishment in Medieval and Renaissance Europe.* Chicago, 1999.

Meyerson, Mark D. *A Jewish Renaissance in Fifteenth-Century Spain.* Princeton, 2004.

Meyerson, Mark D., Daniel Thiery, and Oren Falk, eds. *'A Great Effusion of Blood'? Interpreting Medieval Violence.* Toronto, 2004.

Meyvaert, Paul. " 'Rainaldus est malus scriptor Francigenus'—Voicing National Antipathy in the Middle Ages." *Speculum* 66 (1991): 743–64.

Miller, William Ian. *Bloodtaking and Peacemaking.* Chicago, 1990.

Mitchell, Kathleen, and Ian N. Wood, eds. *The World of Gregory of Tours.* Leiden, 2002.

Mollat, Michel, and Phillipe Wolff. *The Popular Revolutions of the Late Middle Ages.* London, 1973.

Moore, Robert I. "Anti-Semitism and the Birth of Europe." In Diana Wood, ed., *Christianity and Judaism.* Oxford, 1992. 33–57.

———. *The First European Revolution, c. 950–1215.* Oxford, 2000.

———. *The Formation of a Persecuting Society: Power and Deviance in Western Europe, 950–1250.* Oxford, 1987.

Morris, Colin. *The Discovery of the Individual: 1050–1200.* New York, 1973.

Morrison, Karl F. *Conversion and Text: The Cases of Augustine of Hippo, Herman-Judah, and Constantine Tsatsos.* Charlottesville, 1992.

Mundill, Robin R. *England's Jewish Solution: Experiment and Expulsion, 1262–1290.* Cambridge, 1998.

Myers, David N. *Re-Inventing the Jewish Past: European Jewish Intellectuals and the Zionist Return to History.* New York, 1995.

Nederman, Cary J. "Introduction: Discourses and Contexts of Tolerance in Medieval Europe." In John Christian Laursen and Cary J. Nederman, eds., *Beyond the Persecuting Society: Religious Toleration Before the Enlightenment.* Philadelphia, 1998. 13–25.

———. *Worlds of Difference: European Discourses of Toleration c. 1110–c.1550.* University Park, PA, 2000.

Nederman, Cary J., and John Christian Laursen, eds. *Difference and Dissent: Theories of Toleration in Medieval and Early Modern Europe.* New York, 1996.

Nelson, Janet. "On the Limits of the Carolingian Renaissance." In Janet Nelson, ed., *Politics and Ritual in Early Medieval Europe.* London, 1986. 63–67.

Nirenberg, David. *Communities of Violence: Persecution of Minorities in the Middle Ages.* Princeton, 1996.

176

Novikoff, Alex. "Between Tolerance and Intolerance in Medieval Spain: An Historiographical Enigma." *Medieval Encounters* 11 (2005): 7–36.

Ottonian Germany: The Chronicon of Thietmar of Merseburg. Translated by David Warner. Manchester, 2000.

Pakter, Walter. "Early Western Church Law and the Jews." In Harold W. Attridge and Gohei Hata, eds., *Eusebius, Christianity, and Judaism.* Detroit, 1992. 714–35.

———. *Medieval Canon Law and the Jews.* Ebelsbach, 1988.

Pelteret, David A. E. *Slavery in Early Mediaeval England: From the Reign of Alfred until the Twelfth Century.* Woodbridge, Suffolk, UK, 1995.

Peters, Edward. "Jewish History and Gentile Memory: The Expulsion of 1492." *Jewish History* 9 (1995): 9–33.

Pohl, Walter. *Kingdoms of the Empire: the Integration of Barbarians in Late Antiquity.* Leiden, 1997.

Pohl, Walter, and Helmut Reimitz, eds. *Strategies of Distinction: The Construction of Ethnic Communities, 300–800.* Leiden, 1998.

Pseudo-Jerome. *Quaestiones on the Book of Samuel.* Edited by A. Saltman. Leiden, 1977.

Ray, Jonathan. "Beyond Tolerance and Persecution: Reassessing Our Approach to Medieval *Convivencia.*" *Jewish Social Studies* 11 (winter 2005): 1–18.

Ries, Rotraud. "German Territorial Princes and the Jews." In R. Po-Chia Hsia and Hartmut Lehmann, eds., *In and Out of the Ghetto: Jewish-Gentile Relations in Late Medieval and Early Modern Germany.* Washington, D.C., 1995. 215–46.

Rohrbacher, Stefan. "The Charge of Deicide. An Anti-Jewish Motif in Medieval Christian Art." *Journal of Medieval History* 17 (1991): 297–321.

Rosenwein, Barbara, ed. *Anger's Past: The Social Uses of an Emotion in the Middle Ages.* Ithaca, N.Y., 1998.

Roth, Norman. "Bishops and Jews in the Middle Ages." *Catholic Historical Review* 80 (1994): 1–17.

———. *Jews, Visigoths, and Muslims in Medieval Spain.* Leiden, 1994.

———, ed. *Medieval Jewish Civilization: An Encyclopedia.* New York, 2003.

Rubin, Miri. *Gentile Tales: The Narrative Assault on Late Medieval Jews.* New Haven, 1999.

———. "Imagining the Jew: The Late Medieval Eucharistic Discourse." In R. Po-Chia Hsia and Hartmut Lehmann, eds., *In and Out of the Ghetto:*

Jewish-Gentile Relations in Late Medieval and Early Modern Germany. Washington, D.C., 1995. 177–208.

Rubenstein, Jay. *Guibert of Nogent: Portrait of a Medieval Mind.* New York, 2002.

Ruderman, David. "Cecil Roth, Historian of Italian Jewry: A Reassessment." In David N. Myers and David B. Ruderman, eds., *The Jewish Past Revisited: Reflections on Modern Jewish Historians.* New Haven, 1998. 128–43.

Schäfer, Peter. *Judeophobia: Attitudes toward the Jews in the Ancient World.* Cambridge, 1997.

Schmitt, Jean-Claude. *La Conversion d'Hermann le Juif: Autobiographie, Histoire et Fiction.* Paris, 2003.

Schorsch, Ismar. *From Text to Context: The Turn to History in Modern Judaism.* Hanover, NH, 1994.

Schreckenberg, Heinz. "Die patristiche Adversus-Judaeos Thematik im Spiegel der karolingischen Kunst." *Bijdragen, tijdschrift voor filosofie en theologie* 49 (1988): 119–38.

Schwartz, Seth. *Imperialism and Jewish Society, 200 B.C.E. to 640 C.E.* Princeton, 2001.

Scribner, Bob. "Preconditions of Tolerance and Intolerance in Sixteenth-Century Germany." In Ole Peter Grell and Bob Scribner, eds., *Tolerance and Intolerance in the European Reformation.* Cambridge, 1996. 32–47.

Shatzmiller, Joseph. "Jews, Pilgrimage, and the Christian Cult of Saints." In Alexander Callander Murray, ed., *After Rome's Fall: Narrators and Sources of Early Medieval History. Essays Presented to Walter Goffart.* Toronto, 1998. 337–48.

———. *Shylock Reconsidered: Jews, Moneylending, and Medieval Society.* Berkeley, 1990.

Shepkaru, Shmuel. "To Die for God: Martyrs' Heaven in Hebrew and Latin Crusade Narratives." *Speculum* 77 (2002): 311–41.

Siegmund, Stefanie B. *The Medici State and the Ghetto of Florence: The Construction of an Early Modern Jewish Community.* Stanford, 2006.

Signer, Michael A., and John Van Engen, eds. *Jews and Christians in Twelfth-Century Europe.* Notre Dame, IN, 2001.

Simonsohn, Shlomo. *Apostolic See and the Jews.* 8 vols. Toronto, 1991.

Skinner, Patricia, ed. *Jews in Medieval Britain: Historical, Literary, and Archaeological Perspectives.* Rochester, NY, 2003.

Smail, Daniel Lord. *The Consumption of Justice: Emotions, Publicity, and Legal Culture in Marseille, 1264–1423.* Ithaca, NY, 2003.

———. "Hatred as a Social Institution in Late-Medieval Society." *Speculum* 76 (2001): 90–126.

Smalley, Beryl. *The Study of the Bible in the Middle Ages.* Oxford, 1983.

Southern, Richard. *The Making of the Middle Ages.* New Haven, 1959.

———. *Western Views of Islam in the Middle Ages.* Cambridge, MA, 1962.

Stacey, Robert. "The Conversion of Jews to Christianity in Thirteenth-Century England." *Speculum* 67 (1992): 263–83.

———. "Crusades, Martyrdoms, and the Jews of Norman England, 1096–1190." In Alfred Haverkamp, ed., *Juden und Christen im Zeitalter der Kreuzzüge.* Sigmaringen, 1999. 233–52.

———. "Parliamentary Negotiation and the Expulsion of the Jews from England." In Michael Prestwich, R. H. Britnell, and Robin Frame, eds., *Thirteenth Century England VI.* Woodbridge, 1995. 77–101.

Starr-LeBeau, Gretchen D. *In the Shadow of the Virgin: Inquisitors, Friars, and Conversos in Guadalupe, Spain.* Princeton, 2003.

Stocking, Rachel. *Bishops, Councils, and Consensus in the Visigothic Kingdom, 589–633.* Ann Arbor, 2000.

Stow, Kenneth R. *Alienated Minority: The Jews of Medieval Latin Europe.* Cambridge, MA, 1992.

———. "Conversion, Apostasy, and Apprehensiveness: Emicho of Flonheim and the Fear of Jews in the Twelfth Century." *Speculum* 76 (2001): 911–33.

———. *Theater of Acculturation: The Roman Ghetto in the 16th Century.* Seattle, 2001.

Straw, Carole. *Gregory the Great: Perfection in Imperfection.* Berkeley, 1988.

Strayer, Joseph. "France: The Holy Land, the Chosen People, and the Most Christian King." In Strayer, ed., *Medieval Statecraft and the Perspectives of History.* Princeton, 1971. 300–15.

Strickland, Debra Higgs. *Saracens, Demons, and Jews: Making Monsters in Medieval Art.* Princeton, 2003.

Sutcliffe, Adam. *Judaism and Enlightenment.* Cambridge, 2003.

Szittya, Penn R. *The Antifraternal Tradition in Medieval Literature.* Princeton, 1986.

Thomas, Hugh. *The English and the Normans: Ethnic Hostility, Assimilation, and Identity, 1066–c.1220.* Oxford, 2003.

Thompson, E. A. *The Goths in Spain.* Oxford, 1969.

Toch, Michael. "Aspects of Stratification of Early Modern German Jewry: Population History and Village Jews." In R. Po-Chia Hsia and Hartmut

Lehmann, eds., *In and Out of the Ghetto: Jewish-Gentile Relations in Late Medieval and Early Modern Germany.* Washington, D.C., 1995. 77–90.

———. "Economic Activities of German Jews in the 10th to 12th Centuries: Between Historiography and History." In Yom Tov Assis, Jeremy Cohen, Aharon Kedar, Ora Limor, and Michael Toch, eds., *Facing the Cross: The Persecutions of 1096 in History and Historiography.* Jerusalem, 2000. 32–55. [Hebrew].

———. "The Formation of a Diaspora: The Settlement of Jews in the Medieval German *Reich.*" *Aschkenas* 7 (1997): 55–79.

———. "Jews and Commerce: Modern Fancies and Medieval Realities." In S. Cavaciocchi, ed., *Il ruolo economico delle minoranze in Europa. Secc. XIII–XVIII.* Florence, 2000. 43–58.

———. *Die Juden im Mittelalterlichen Reich.* Munich, 1998.

Tolan, John V. *Saracens: Islam in the Medieval European Imagination.* New York, 2002.

———, ed. *Medieval Christian Perceptions of Islam: A Book of Essays.* New York, 1996.

Trautner-Kromann, Hanne. *Shield and Sword: Jewish Polemics against Christianity and the Christians in France and Spain from 1100–1500.* Tübingen, 1993.

Ullmann, Walter. *The Individual and Society in the Middle Ages.* Baltimore, 1966.

Van Dam, Raymond. *Saints and Their Miracles in Late Antique Gaul.* Princeton, 1993.

Van Engen, John. "Christening the Romans." *Traditio* 52 (1997): 1–45.

Webster, Leslie, and Michelle Brown, eds. *The Transformation of the Roman World, AD 400–900.* Berkeley, 1997.

Wickham, Chris. *Early Medieval Italy: Central Power and Local Society 400–1000.* Totowa, NJ, 1981.

Wolff, Phillipe. "The 1391 Pogrom in Spain: Social Crisis or Not?" *Past and Present* 50 (1971): 4–18.

Wolfram, Herwig. *History of the Goths.* Berkeley, 1988.

Wood, Ian. "Pagan Religion and Superstitions East of the Rhine from the Fifth to the Ninth Century." In G. Ausenda, ed., *After Empire: Towards an Ethnology of Europe's Barbarians.* Suffolk, UK, 1995. 253–79.

Yerushalmi, Yosef. "Exile and Expulsion in Jewish History." In Benjamin Gampel, ed., *Crisis and Creativity in the Sephardic World: 1391–1648.* New York, 1997. 3–23.

————. *Zakhor: Jewish History and Jewish Memory.* New York, 1989.

Yuval, Yisrael. "Easter and Passover As Early Jewish-Christian Dialogue." In Paul F. Bradshaw and Lawrence A. Hoffman, eds., *Passover and Easter: Origin and History to Modern Times.* South Bend, Ind., 1999. 98–124.

————. "Jews and Christians in the Middle Ages: Shared Myths, Common Language." In Robert. S. Wistrich, ed., *Demonizing the Other: Antisemitism, Racism, and Xenophobia.* Singapore, 1999. 88–107.

————. " 'They Tell Lies: You Ate the Man': Jewish Reactions to Ritual Murder Accusations." In Anna Sapir Abulafia, ed., *Religious Violence between Christians and Jews: Medieval Roots, Modern Perspectives.* New York, 2002. 86–106.

————. "Vengeance and Damnation, Blood and Defamation." *Zion* 58 (1993): 33–90. [Hebrew].

INDEX

Abraham of Saragossa, 49

Abulafia, Anna Sapir, 93

Adalberon, bishop of Metz, 59

Africa, 41

Agobard, bishop of Lyon, 48–49, 149n17

agriculture, 19; in Carolingian era, 46, 47; in Gregory the Great, 35; in Speyer, 63; in tenth- and eleventh-century Europe, 55, 63; in twelfth-century Europe, 64; in Visigothic Spain, 38. *See also* peasantry; rural areas

Ahimaaz ben Paltiel, *The Chronicle of Ahimaaz*, 56–58

Alan of Lille, 92

Albi, 101

Alsace, 107

Amittai, rabbi (in *Chronicle of Ahimaaz*), 57

Amolo, bishop of Lyon, 49

Anabaptists, 133

Andrew, bishop (in Gregory the Great), 30

Annon, bishop of Cologne, 59

Aquitaine, 101

Arab culture, 56

Aragon, 102, 112

Arians, 13, 17, 21, 29, 36, 37

Aribert, archbishop of Narbonne, 46

Armentarius (in Gregory of Tours), 22–23

Armleder attacks, 106–7, 108

ashkenaz. See hasidei ashkenaz

Augsburg, 106, 107, 131, 132

Augustine, 3, 12, 148n4

Aurasius, bishop of Toledo, 39

Austria, 107, 108

Avitus, bishop of Clermont, 24–25

Bamberg, 131

baptism, 39, 41, 45, 50. *See also* conversion

Barcelona, 111

Baron, Salo, 5

Basel, 131

Basilius the Hebrew (in Gregory the Great), 35

Basnage, Jacques, 3

Bavaria, 105, 131

Benedict of Nursa, 29

Berlin trials, 132

Bernard of Clairvaux, 103

Bernstein, Michael André, 5

Beziers, 101

Beziers, bishop of, 87

Bible, 1, 14–15, 21, 44, 45, 55, 67, 114

bishop(s): and conversion of Herman-Judah, 72–73; in Gregory of Tours, 24; in Gregory the Great, 29–30, 34; patronage of, 36; and Rhineland persecutions, 77; use of relics by, 15; and violence in Germany, 106; in Visigothic Spain, 37. *See also* Catholic Church; clergy; pope/papacy

Blanche of Castille, 101

blood libel accusations, 94, 129, 130. *See also* cannibalism, ritual; host desecra-

commerce/trade: in Carolingian era, 46,
47–48, 49, 50; and charter of Rudiger
of Speyer, 60–61; local authorities' desire for, 59; in medieval Germany, 83;
and Peter II, 50; in Speyer, 63; in
tenth- and eleventh-century Europe,
55; in twelfth-century Europe, 64;
in Visigothic Spain, 41. *See also*
merchants
Constantine, 13
conversion, 150n19; in Carolingian era,
44, 45, 47, 50; and expulsions, 117;
and First Crusade, 79, 80, 81–82;
forced, 12, 13, 21, 24, 29, 31, 39, 40,
41, 93, 96, 113; in Gregory of Tours,
21, 23, 24–25, 27; in Gregory the
Great, 29, 30, 31, 32, 33–34; and
Henry I of Germany, 50; of Herman-
Judah, 72–73; and interior identity,
70, 71; in Italy, 125, 127; and Luther-
anism, 3; in Minorca, 13; and miracles, 71, 72; of New World peoples,
133; and papacy, 93; of Reccared, 36–
37; in Roman law, 12; in Spain, 39–
40, 41, 110, 111, 112, 113, 114, 120–
21; in tenth- and eleventh-century Europe, 55; in thirteenth-century England, 68; in twelfth-century Germany, 68–69; and twelfth-century
Jewish-Christian acculturation, 68–74.
See also persecution(s); violence
convivencia, 75–76, 100, 113, 135–36,
137. *See also* social integration
Counter-Reformation, 94, 126
crime, 90–91
crucifixion, ritual, 96, 103
Crusade(s): against Cathars, 100–101;
and expulsions, 117; First, 5, 11, 54,
66, 71, 76–83, 84, 96, 103, 105, 111,
136; following famine of 1315–17,
101; and martyrdom, 66; and Muslim

threat, 92–93; Second, 96, 103; and
York massacre, 99. *See also* chronicles
cultural integration, 64–74. *See also* social integration
culture, 9, 92
Cyriacus (in Gregory the Great), 32–33

David b. Meshullam, 62
deicide, 6, 92
dependent status, 34, 35. *See also* slavery
diaspora, rhetoric of, 1–2, 3
disputation, 20–21, 67, 93, 110. *See
also* rhetoric
Donatists, 13
Druthmar, 45

economy, 96, 122, 124
education, 44, 45, 57, 66–67
Edward I, King of England, 117
Edward II, King of England, 118
Egica, 41
Einhard, 49
Eliezer bar Nathan Chronicle, 77
Emicho (Crusader), 78
England, 87; class rebellion in, 111;
conversion in, 68; crime in, 90; and
expulsions, 7, 116, 117–18, 120, 121,
132; Norman conquest of, 64; in
post-Carolingian era, 52; settlements
in, 85; social integration in, 75, 84–86,
88, 137; violence in, 85, 97–100, 117;
war in, 91
Enlightenment, 2
Erfurt, 103, 109
Erwig, 41
ethical monotheism, 2
ethnicity, 4, 27, 28, 95
Eunomius (in Gregory of Tours), 22
exile, 1, 27–28, 55
expulsion(s), 8, 10, 76, 115, 116–22; as
ad hoc solutions, 137; after 1290, 7;
and England, 7, 116, 117–18, 120,

Milton Keynes UK
Ingram Content Group UK Ltd.
UKHW011719260424
441804UK00007B/169